Don't Over React!

Don't Over React!
An entertaining guide to help you succeed
in college-level general chemistry
First Semester
Eric G. Chesloff

ISBN: 9798671395297
Imprint: Independently published

Published by Eric G. Chesloff, near Philadelphia, PA
Contact: ecdc5@verizon.net

Dedication

To my darling wife, Janis, for her constant and unwavering love, encouragement, and support throughout this project.

Contents

Acknowledgements

I thank my gracious Lord and Savior Jesus Christ, who patiently continues to guide me.

Thank you to my wonderful wife Janis, who read the manuscript and provided valuable editorial input.

Thanks and appreciation are due to Dr. Richard Schwartz, who read the manuscript and provided comments.

Thanks are also due to Katherine Kama'ema'e Smith of Honu Media LLC, Research and Writing Services, for valuable editorial input and encouragement.

Finally, thanks to my publisher, David Bergsland and Skilled Workman Publications, who patiently endured all the changes of the rough cut manuscript en route to a polished gem.

Introduction

I believe that most undergraduate students take general chemistry because they must; their course of study requires it. For this reason in many cases this causes them great distress. You might be part of this group. You're experiencing emotions from depression to outright fear. You want nothing more than to pass the course with a C if possible and get on with your education and your life. Some students have an easy time with the course, but unfortunately many have an experience far different. The latter may have preconceived notions they have either created or allowed to be created by others. That's why I've written this companion book. Let me explain.

There's a plethora of general chemistry textbooks out there already, each no doubt with a rather high price tag. So the very last thing we need is another blasted textbook composed in agonizingly clinical, pedantic, and patronizing detail. Then, there are the so-called "dumbed down" chemistry books, which in my opinion are textbooks masquerading as primers. They might be even more patronizing!

Therefore, I did **NOT** want to re-invent the wheel. My goal was to publish a monograph of simplified and streamlined chemistry with some commentary and exhortation, one that readers would find more than palatable. My desire is that you would not only survive chemistry, but actually thrive and perhaps take a liking to it! Wouldn't that be astounding?!

Instead of "lecturing" to you, as the language of textbooks typically does, I want to have a chat with you, a sort of conversation. Of course, it will be a monologue rather than a dialogue, but that's the style I want to use. I want you to be comfortable using this book. Of course, you can react (no pun intended), but I won't be there to respond.

Having said all that, don't throw out your rather expensive textbook and don't try to re-sell it. This book is a ***companion***, **not a replacement or substitute**. Hopefully the material will clarify and/or embellish what's inside that textbook. Remember, you are primarily bound to follow your instructor's requirements for the course. But I want to make chemistry so easy to understand that you'll yell something like *"I am simply in love with chemistry,"* or *"I'm going to name my first-born after Dr. Chesloff" (yes, yes, Erica, if necessary).*

I think everyone should take at least one chemistry course *(Right Dr. C., that's like saying I think everyone should jump into an alligator pond just once!)*, not because I'm a nerd (which I've been called more than once), but because it's a basic, physical science with pervasive application, and knowing some chemistry allows you to make tremendously more informed choices in the world. Actually, it doesn't matter what field you decide to pursue, whether science, engineering, business, or other humanities fields. **Learning how to think scientifically is ALWAYS useful.** If you learn to think scientifically, you can apply it to **virtually any field.**

I truly wish you well in this course and beyond it. Later, I would appreciate your feedback, if you're so inclined, whether positive or negative (😄). It would be truly heartwarming to hear that this book has helped you, and/or that there are constructive ways to improve it.

Thank you very much.

EGC Summer 2020

Study Methods

Some students glide effortlessly through chemistry, but most probably require a more disciplined approach. Below I offer several ideas that can help you succeed in the course.

- Find out asap about the campus/departmental tutoring program and avail yourself of it sooner rather than later, if necessary.

- Study **every day**, and not for hours at a time necessarily, but take some time to read AND to write. That's right; write out notes and problems. Don't just skim or read your notes. Interact! There is something about actually writing that establishes that hand-head connection and later reinforces learning. Do not neglect this point.

- Study with one or more other students. The effectiveness of peer to peer teaching cannot be overstated.

- At the end of each chapter of your textbook, the author includes a summary outline of the major (and minor) points. **Write your own outline!** Every student has a unique way of learning information. Writing your own outline enables you to reinforce your learning by triggering key concepts in the mind from your own writing style, including symbology and phrasing.

- Consult general chemistry textbooks besides that of your author. Not to demean your author's (or lecture instructor's, for that matter) teaching style, but if your textbook does not provide adequate explanations to clarify a topic, others probably will. Don't hesitate on this one. In research, during experimental design, scientists typically consult a wide variety of sources when conducting a literature search.

Using This Book and
My Approach to Learning

Of course, for basic concept edification, you can read this book by itself. However, I believe it will be most useful to you if you **have it available while you are studying from your textbook and notes.** Then you can immediately turn to it for reference and clarification as necessary. I have deliberately omitted the details and exhaustive explanations incumbent on a textbook author to include. My goal, as I stated, is to present concept and application to you very plainly and in a streamlined manner so as to make chemistry more readily understandable and practicable.

My former students could attest that I encourage a holistic approach to chemistry. I believe that besides deductive reasoning and quantifying, which of course are essential to succeeding in science, one needs to recruit or cultivate a sense of artistry and imagery. Periodically, stop and visualize a collection of molecules within various scenarios. This collection typically runs into the billions and trillions. You don't just deal with one or two molecules at a time. This is especially true, therefore, with subatomic particles such as protons and electrons. Periodically you should pause (and I may ask you) to imagine certain molecular arrangements in terms of natural symmetry. I am convinced that you need to use both sides of the brain ("left brain/right brain") to fully comprehend chemistry.

Chemical Topics

Traditionally, all chemistry professors start out by introducing students to the ***scientific method***. This may shock you, but you're already using it! I don't mean in the classroom, but outside in the real world. When you go shopping, whether online or in a brick and mortar store, you do research, compare prices, investigate brands, note the length of a sale, compare coupons, etc.

Mrs. C. is an excellent researcher in that regard. She'd make a great research chemist…Oops. Don't tell her I said that. She hates chemistry.

Anyway, there are essentially three parts to the method:

> ***General observations*** of your world that lead to a question or problem, known as the ***hypothesis***,

> ***Investigating*** that problem or question, which involves making observations and collecting data, called ***experimentation,*** and coming to a

> ***Conclusion***. Then we figure out the next direction to take, based on the evidence we gathered. Then the loop repeats. Eventually we gather enough evidence to develop a fairly firm **explanation** of our **observations**, called a **theory**.

DO NOT CONFUSE A ***LAW* (the "scene") with a *THEORY* (behind the "scene")**! A ***law*** is a **summary statement** of many observations leading to the same conclusion, i.e., whenever A happens, then B follows…of course, as long as B does follow, invariably. What goes up comes down. Whenever it rains for more than an hour, the basement floods. It's invariable, irrefutable, immutable. You don't argue.

Don't I sound so sophisticated when I use these big words?

You get the idea. You get the same result each time, and you form a **summary statement** of those results as stated above.

Just so I know you understand, show me by giving me an example from your own life experiences. I'll start you off: "Whenever_______________________________________, _______

______________________________________.

Now, a **theory** *explains* what we see, sort of like behind the scenes. For example, when we toss a small piece of sodium metal into water, we get a vigorous reaction easily observed. That's the **law** part: when sodium hits water, the reaction is vigorous and complete! Of course, the statement of the law may be much more general. The **theory** part is, when sodium atoms meet water molecules, the sodium atoms' outer electrons are readily driven to the water molecules in order for sodium to achieve a more stable noble gas configuration (that of neon).

Get the difference? Here's an analogy: You watch a magician perform a fascinating trick, and he obligingly repeats it several times with the same result. That's the law part. The real mystery is "how does he do it?" That's the theory part, what you don't see.............

OK Dr. C., don't lay it on so heavily!

Well, no apologies; too many students get so confused over the differences! Now, see if you can illustrate your understanding of *law* and *theory* using your own examples!

Law: ____

____

Theory: ____

____

Law: ____

____

Theory: ____

____

Pretty much all of us, in nearly every profession, use the scientific method: a detective in crime investigation, a lawyer in gathering evidence and examining witnesses, a building engineer in evaluating project feasibility and availability and cost of building materials, a chef developing novel recipes. It's indispensable, really. You're going to use the method rigorously in your chem course.

By the way, even if you never take another chemistry course *(Dr. C., you got that right!)*, the scientific way of processing information will help you IN ANY FIELD YOU END UP! That's right, no exaggeration! You will need to think critically in medicine, law, engineering, or in business. I know many of you aspire to open your own businesses. I don't care what it is (no, no, of course I CARE, but what I'm saying is that the kind of pursuit is immaterial). I always want my students to achieve the highest success in whatever direction they choose to go. Thinking scientifically may not only help you succeed, but may very well "save your life" in ways you might not imagine now.

Problem Solving

Throughout this text, I'm going to encourage you to:

1. Think practically: If the concept or principle has no practical application, what good is it? When I teach, I search for practical type problems.

2. During problem solving, FIRST SET THE PROBLEM UP AND ESTIMATE THE ANSWER! You will be training yourself to analyze the problem and, in the process, avoid wrong answers. You will also likely earn at least part credit for an open-ended problem.

3. Examine the answer or result. Is it reasonable? For heaven's sake, NEVER submit an answer or result to ANYONE without deciding that it's REASONABLE!

I may periodically (no pun intended) make some outrageous-sounding statements, but I challenge you to think about them. I unapologetically challenge my students to "think outside the box" and solve unconventional problems. Remember, many scientific and business revolutionaries were initially ridiculed for their "preposterous" ideas.

The Method

In class and textbook, you'll be introduced to the ***factor label method*** for problem solving. In essence, this refers to stringing together all the numerical components of the problem so that it's **set up** to solve in one continuous operation by calculator. The key here is to examine the dimensional units of each component and arrange the problem so that the units divide out with the exception of the one(s) we want in the answer. This is called, appropriately, ***dimensional analysis***. Here's an example:

Convert 3.65 cm to inches.

OK, we need a conversion factor that will take us from cm to inches. There are 2.54 cm per inch. Now, we need to arrange the units so that the cm ***cancel out*** leaving us with inches.

cm x inch/cm = inches (~~cm~~ x inch/~~cm~~)

Arrange the units first, then insert the numbers.

3.65 cm x 1 inch/2.54 cm = 1.44 inch

But can't I just use cross-multiplication as I did in high school?

Certainly you can, and I permit my students to use it if they insist. But it's much more efficient to use the factor label method, particularly with more complex problems that will come later.

Try this one:

A certain type of carpet sells for $12 per square yard. I want to carpet my 20 ft. x 10 ft. living room with it. What would be the total cost?

Ugh, you're getting into two dimensions. Bad news!

Hold on; it's not so bad. First, create a simple map:
square feet → square yards → cost ($). Are you with me?

Here we go:

20 ft. x 10 ft. = 200 ft²

$$200 \text{ ft}^2 \times \frac{(1 \text{ yd})^2}{(3 \text{ ft.})^2} \times \frac{\$12}{1 \text{ yd}^2} = \$267$$ (rounded to dollars; if we were stricter in following sig fig rules, the answer would be reported as $270)

Did you follow the route? This is a good teaching problem, because it shows the use of the factor label method in general, and a multidimensional problem in particular. When converting square or cubic anything, revert first to the **single dimensions** (i.e., ft., yds. etc.), express that as a conversion ratio (i.e., 3 ft. /1 yd.), encase the **entire** expression(s) in parentheses, and **then clear** the parentheses.

With clearing of parentheses, the above problem becomes

$$200 \text{ ft}^2 \times \frac{1 \text{ yd}^2}{9 \text{ ft.}^2} \times \frac{\$12}{1 \text{yd}^2} = \$267 \text{ (etc.)}$$

The numerical answer is actually a moot point, at least in the classroom. Don't misunderstand; your instructor/colleagues/employer will need to know that number accurately. What I stress in the classroom is primarily setting up the problem as above. The answer can be easily obtained once that's properly done.

As stated above, **pre-estimate the answer** before actually calculating, just as you do in the real world. Admit it; you estimate your cost at the gas pump and at the register at the supermarket, BEFORE the purchases are actually totaled up. When they finally are, they should sound **reasonable**. Ditto in the chemistry classroom. Everyone makes calculator errors, and in the real world we want to avoid being overcharged. In some cases wrong calculations could be tragic.

Mrs. C. is sharp as a tack. Dining out, she's caught servers' errors more times than I can recall. She'd be an excellent private ◉!

Modern Chemistry: Beginnings

Now for some history, your favorite! It's easy to argue that chemistry is a fairly young science, whatever that means. The ancient Greeks gave us quite a bit in terms of academia. In particular, they also thought about matter itself, and gave us the term "**atomos**," which means uncuttable.

They also gave us feta cheese, which IS cuttable (sorry, couldn't resist it- 😄)

They thought matter could be reduced ultimately to uncuttable spheres, which were its building blocks. They weren't totally wrong; you can deconstruct the atom, but it takes energy, and, depending on your goal, more or less of it. Unfortunately, this rudimentary atomic theory was ignored for the next 2,000 years. In the 1700s, there was quite a bit of experimentation going on. Back in those days, all folks could do was weigh some of each substance and see how much weight of another substance would react with it. However, they were able to draw some important preliminary conclusions about the behavior of matter.

We really don't hit the first major intersection until 1804, the year generally considered to be of the birth of modern chemistry. That's when John Dalton, a schoolteacher, summarized the myriad prior observations of the behavior of matter into three (or four, depending on who's authoring) postulates collectively called the modern **Atomic Theory**.

By the way, a lot of those early folks were into other professions and occupations before or while they delved into chemical investigation. Since everything was so simple back then, one could afford to do that. Try moonlighting today as a chemist (😄)! That's like moonlighting as a brain surgeon…………

The Dalton Dance

What the **Atomic Theory** says essentially is that atoms are the uncuttable, rock bottom building blocks that make up everything, kind of like what the ancient Greeks thought. They combine in whole number ratios! Also, all atoms of a given element are alike; all iron atoms are the same, all sulfur atoms are the same, etc. And finally, a chemical reaction is nothing more than atoms getting unglued from each other and shifting around to get glued onto new neighbors to create new substances. It's nothing but a huge dance with atoms changing partners. If you understand that, you're 90% over the basics.

Dalton's thinking was actually revolutionary, notwithstanding some minor (😄) adjustments that came later. For example, atoms are not uncuttable; behold **protons**, **neutrons,** and **electrons**, to start! Also, he did mention that all atoms of a given element (i.e., carbon) are alike. But we know now that each element comes as a series of **isotopes**, which means each type has a nucleus with different numbers of neutrons. I won't elaborate on this here, but consult your textbook or, if you have time, google "isotopes." By the way, the number under the symbol of each element in your Periodic Table is the ***weighted average mass***, taking into account the mass and population of each of its isotopes!

Besides those variations, Dalton's theory is still basically rock-solid. That's pretty darned good for a theory over two hundred years old!

I mentioned protons, neutrons, and electrons. Between the late 1800s to early 1900s, we discovered that atoms contain a nucleus (sort of like a biological cell nucleus), made up of protons **(+1 charge)** and neutrons **(0 charge)**. Protons and neutrons have approximately the same masses, neutrons with a slight edge. Together they make up essentially the **mass** (weight) of the atom. The third particle, the electron **(-1 charge)**, defines with its fellow electrons the **volume** of the atom.

Electrons weigh about 2000 times less than either protons or neutrons, so on a relative scale they really don't weigh anything. Of course they have mass, but it's negligible. For example, we'll find out that atoms and their ions (look it up) are considered to weigh essentially the same. Atomic sodium (Na) and ionic sodium (Na+) each weigh essentially 22.99 grams per mole (look it up).

The **Atomic Theory** is undergirded by the **Law of Conservation of Matter**, the **Law of Definite Proportions**, and the **Law of Multiple Proportions**. You can look these up, but I'll state them in my own words, as you should learn to do!

The Law of Conservation: We can only create new matter from the matter we already had, and both sides of the chemical equation must balance.

I'll illustrate the **Law of Definite Proportions** (LDP) and the **Law of Multiple Proportions** (LMP) as follows:

LDP: Water will always be composed of one atom of oxygen and two atoms of hydrogen in a ratio of 1/2 or 2/1 (depending on point of view, 😄), and

LMP: When comparing carbon monoxide to carbon dioxide, given the fixed amount of carbon (one atom), the oxygen atom ratio between the two compounds is 1/2.

The key point here is that the ratios are of WHOLE NUMBERS! Remember this: **Dalton was a whole number guy**.

I generally discourage memorizing definitions; it doesn't confer any real learning. Oh, and it's excruciatingly boring! Of course, I'm never boring; both of my friends will confirm that..........

Now, quickly, in the allotted spaces below, give your own examples of these three Laws stated above. **NO DEFINITIONS! I'm watching you!**

1)

2)

By the way, if you examine these three laws again, and then the **Atomic Theory**, you can see how the theory "explains" the Laws and gives us behind the scenes views at atomic level. Are you more comfortable now with the difference between **law** and **theory**?

Now, I mentioned the period between the late 1800s and early 1900s. During that interval of time we established the existence of the nuclear atom. I won't elaborate on the discoveries that led to this, but I'm alerting you to be able to summarize **IN YOUR OWN WORDS** the experiments and conclusions of: **J. J. Thomson, R. A. Millikan, Ernest Rutherford,** and **James Chadwick.**

You also should probably become informed about **alpha particles** (doubly charged helium nuclei), **beta particles** (high-speed electrons), and **gamma rays** (neutral rays, with energy a bit above that of x-rays) to aid your overall understanding of that part of chemical history. There's probably some mention in your textbook of **Marie Curie, Henri Becquerel,** and possibly **Wilhelm Roentgen**. I'd take note of these scientists' chemical contributions. Google them if you have time.

Here's an outrageous idea: Why do many reactions just happen in the first place? The answer: things must be more stable on the other side. Nature seeks stability. Chemical processes, at least those that seem to readily go with just a little push, don't happen to make things less stable than they were starting out. This is not a glib response. In the vast majority of reactions you will encounter, they most likely go to completion on their own, because something more stable is waiting to form. Otherwise, they simply wouldn't happen! Imagine that. Here's an analogy: Skiers generally head down the slope spontaneously rather than up it.

If you happen to see (or think you see) the latter, let me know, and I'll arrange psychiatric evaluations for BOTH of us!

The Kingdom of Matter

Just as in biology, chemists have a "kingdom" called ***matter***, which is really the physical stuff of the universe. It's subdivided into **substances** and **mixtures.** Some examples of substances: sugar, iron, table salt, water. Now, what do all those have in common? Hint: "pure." If, for example, I take a sample of sugar, I expect it to be _______________________. Anyway, I will let you, rather than myself, make a summary statement about substances right here, IN YOUR OWN WORDS.

____

_______________________________.

Now, can you tell me the difference between water and iron (pure iron, that is)? Right; iron is one **element** through and through (#26 on your Periodic Table), whereas water has oxygen (element #8) and hydrogen (element #1), two different elements, glued together in some fashion. So we'll say that iron is an ***element***, and water is a ***compound***. Elements are the simplest kind of matter chemically, as we said earlier, but compounds have at least two different elements connected as stated above. Either way, they're both ***substances.***

Here's the really fun part. Elements and compounds can combine together and rearrange to form totally new substances (remember the **Dalton Dance**?).

BE READY TO GIVE YOUR INSTRUCTOR EXAMPLES OF A SUBSTANCE (ELEMENT AND COMPOUND) AND OF A MIXTURE (HOMOGENEOUS AND HETEROGENEOUS).

Memorize definitions if you must, but giving examples demonstrates true understanding.

By the way, what about the terms **Matter** and **Mass**? What's the difference? Your textbook says that matter is anything that occupies space and has mass. But the "street" definition is better: Matter is the physical "stuff" of the universe, as I mentioned above. Isn't that easier to relate to? I know, I know, it doesn't sound technical. But the one I'm giving you is rock-bottom and the most useful.

Now, what is mass? Book definition: The actual quantity of matter. But how do we define the quantity? Loosely speaking, it's weight. No, no, you say. They're not the same. The actual quantity of matter doesn't change, whether you're on the Earth, or the Moon, or Mars. But the weights will change, as you remember from grade school, due to gravity differences. I counter, you're probably going to stay on the Earth for the entire semester.

Actually, given the recent advancements in space travel, some of you just might be leaving the Earth for a while. Please let me know if you're planning on it! I'll probably respond with "have a safe trip," or "don't forget to journal!"

I've actually heard somebody say, "Go mass this on the balance." Hilarious, isn't it? How much do you mass? Yuk, yuk. The joke is on them, really. 😆

Then we have # volume. You probably understand this already, but in case you don't, how much space are you taking up right now? That's your volume. It's different from mass (weight), because we're measuring the product of length, width, and height to get it.

Some useful relationships: One cubic centimeter (cm^3, cc) is essentially equal to one milliliter (mL). Also, in the lab, a useful approximation for one mL is 20 drops.

Now, if you divide mass by volume, you get *density,* typically measured in grams/milliliter (g/mL). If we have a regular solid, we can determine that as grams per cubic centimeter, typically (g/cm^3). Density is what's called an *intensive* property, because it's constant regardless of the size of the sample. Guess what the opposite property is called.

By the way, if you're taking the lab portion of the course, keep in mind that there are essentially two operations in the lab: *pouring* and *weighing*. Everything else is just reading the marks and between the marks on the glassware or reading the digits on the balance you're using. Then it's calculate, calculate, calculate!

Now, without wasting time, I'll say this only once: **KNOW YOUR GREEK PREFIXES, KNOW YOUR GREEK PREFIXES, KNOW YOUR GREEK PREFIXES.**

So, Dr. C., exactly which ones do I need to know?

It's arbitrary, based on what your lecture instructor wants. Suffice to say **there's no excuse for not knowing them!**

There; that should get me off the hook!

The Data Thing

If you take chemistry lecture, you most likely are taking lab along with it. So you're going to have to write a lab report sooner or later, and you'll need to show how to do chemical calculations and properly report the answers. That means you need to know something about significant figures ("sig figs"). Your lecture instructor may be lenient about this on quizzes and exams, but expect a more stringent attitude from your lab instructor.

In order to understand the principle of sig figs, we need to first discuss **precision**. The term "**precision**" (the fineness or exactness) fundamentally refers to the sensitivity of a piece of glassware or balance in terms of giving you a measurement reading.

Practically, **precision** refers to the number of **visible divisions** (marks) on the instrument/glassware. The more marks, the greater the precision. For example, **beakers are generally very imprecise, and we don't use them but for rather general capacities**. Burets, graduated cylinders, volumetric flasks, and pipets, on the other hand, are rather precise, and we can measure typically within parts of a milliliter (mL) with any of them. Get the idea?

Depending on the requirements of the experiment, the type of glassware or other instrument (i.e., balance) will limit the precision of your reported result(s). You simply can't go beyond those limits, right? If that makes sense, you're most likely to "get" sig figs right off the bat.

You **ARE** accountable to report in lecture and lab the proper numbers of sig figs in data and calculations. As the saying goes, remember to "cross your t's and dot your i's."

Actually, what if you rather crossed your i's and dotted your t's? You'd still end up with a t and an i, right? Whatever.......

Your textbook does a thorough (spelled B O R I N G) job describing sig figs. I say, let's jump in and do some examples. By the way, I've found that everybody learns them soon enough.

How many sig figs do the following numbers have?

2436 _____

0.00305 _____

0.08640000 ______

Answers: 4, 3, & 7, respectively. If you got them all, you're batting 1000 (*one sig fig,* 😄), or more precisely 1.000 *(four sig figs in baseball math).*

OK, NOW you're allowed to look at the sig fig rules in your textbook.

By now you are aware of the two parameters in data treatment: **accuracy** and **precision** (relax, it's in a bit different context below). With **accuracy**, you need a **reference value**, or a universally known and accepted numerical goal to shoot for. **By the way, you will see the terms <u>true value</u>, <u>accepted value</u>, or <u>standard value</u> in place of <u>reference value</u>. File these terms away; they all mean the same thing: the target.** Sometimes you are given the reference value before an experiment, and sometimes after (as with unknowns). The goal, of course, is to hit the target.

Precision, on the other hand, implies that you've done several trials of the experiment and the results all pretty much agree with each other. But it doesn't necessarily mean you've hit the target. If your precision is good but accuracy is bad (way off target), there could be a systematic error that's being repeated, most likely involving instrument calibration but, possibly, your technique. It's a good idea to make sure all instruments are properly calibrated daily AND that you're reading your glassware properly before you record anything.

Generally, it turns out that if your precision is good, it's rather likely that your accuracy is good also.

With lab reports, you're going to have to

1) show a sample calculation, which by the way usually involves mainly multiplication/division, and

2) report the end result with the **proper number of sig figs.** I know; I said that already. Follow the rules, coupled with an understanding of accuracy and precision (see above and below). **DON'T GET LAZY WITH THIS!** You are sure to lose points on the lab report if you do.

What's the big deal about sig figs anyway? They're a royal pain!

I know, I know, but here's the deal. Just a reminder: You can't take liberties with the precision of a measurement **if the measuring device won't let you do it**. It's a flight of fancy that's actually rather inappropriate and in many cases will challenge your credibility in the real world.

The Atom and Ion Shuffle

So, the neutral atom contains a nucleus which is positive, composed of protons and neutrons, and it has electrons, which are negative. A neutral atom has the same number of electrons as protons. Electrons whirl about the nucleus kind of like planets around the sun at a sizeable fraction of the speed of light. As I mentioned earlier, the nucleus composes the essential **mass** of the atom, and the electrons establish the **volume** of the atom.

So where's the chemistry? Really, we are going to move electrons from atom to atom. Sometimes, the electrons will be completely donated from one atom to another, and sometimes shared, like community property.

Ions are simply charged single atoms or clusters of atoms. Examples are Cl-, Na+, ClO_3^-, SO_4^{2-}. Positive ions are called **cations**, and negative ones **anions**. If you think about it, if the electrons are the only particles that move, as I mentioned above, the atom or cluster of atoms with a net negative charge has acquired extra electrons, and the cluster that has a net positive charge must have lost electrons somehow.

Someone recently told me this joke:

A: "Oh no! I've lost all my electrons!"

B: "Really? Are you sure?"

A: "Yes, I'm positive!"

You had better at least be smiling!

(Yes, like that!)

OK, so now we have these positive and negative ions. What do we do with them? No doubt your text devotes quite a bit of space describing **ionic compounds**, which are simply compounds made up of ions! They commonly go by the generic name "salts."

So, how did they become ions in the first place?

That's not our concern right now. All you need to know is that your instructor told you to memorize the charges and names of the ions, so focus on doing that.

In Appendix I there's some relevant organizational material to lessen that labor. In the end, though, I just tell my students to learn them. I hate having to memorize and I know you do as well. Most of the time I discourage it in favor of learning the concept. However, in some cases, due to scope and time constraints, it's unavoidable. Whatever, keep in mind that the more you use these ions, the better you will know them.

It's agonizing to me that students too often fail to recognize ions, particularly polyatomic ones, within ionic compounds,

20

so they make horrific mistakes trying to name the compounds. I am therefore on my knees begging you to learn to RECOGNIZE POLYATOMIC IONS WITHIN COMPOUNDS, not merely alone!

The Electrical Neutrality Principle

There's a very simple principle you learned in grade school. Opposites attract. That means that positive particles are attracted to negative ones. They naturally want to neutralize each other. It's as simple as that! **So, however many positive charges we have, that's how many negative ones we'll need to neutralize.**

Quite literally, the rest is math. Therefore, if sodium ion is 1+ and sulfide ion is 2-, how do we create a neutral situation using the least number of ions? I hope the answer is obvious; we need two sodium ions for every sulfide ion. That's how we arrive at Na_2S as the formula for sodium sulfide.

Terms like "least common multiple" might be running through your head right now, and that's fine. Just fall back on the ***electrical neutrality*** principle and you can't go wrong. I will say no more on this; just practice, since there are of course many possible combinations. Again, see Appendix I at the end of the book about polyatomic structure and naming.

You most likely will encounter molecular compounds with only two different elements (or **"binary"** compounds). Both will likely be non-metals (upper right section of the Table). The first element typically lies further to the left and downward in the Periodic Table versus the second element. The rules for this are simple: Use the Greek prefixes, and keep the first name intact (i.e., *di*nitrogen, *tri*phosphorus...) and drop the ending of the second and substitute _ide.

Naming Compounds

Rather than a protracted discourse on compound naming, I've included a page on formula writing and naming at the end of the book in Appendix I.

Yeah, right, Dr. C.; you're just trying to blow it off!

Not at all. Your instructor will supply the suffixes, etc., and then probably turn you loose to memorize them. Bottom line.

Writing and Balancing Equations: No thumbs on the scales at THIS deli counter!

A chemical equation is very much like one in math, but instead of the "=" sign, we have an arrow ($\rightarrow$). The arrow signifies that a dynamic process is occurring over time, and that certain substances on the left side, the **reactants**, are combining and rearranging to form the substances on the right side, the **products**. The Dalton Dance!

The principle behind the equation: The Law of Conservation of Matter dictates that the total number of atoms or mass on the left must be the same as the total number of atoms or mass on the right!

In order to properly write a chemical equation, you must first be able to write chemical formulas. Otherwise, you will write an erroneous equation, and therefore will NEVER BE ABLE TO PROPERLY BALANCE IT. This is tragic and may actually be dangerous in some cases! Remember, I said something earlier about potentially tragic results with erroneous calculations, and you will see how these are connected!

I do suggest to students **two rules** when balancing.

1) The ***odd/even*** rule: Make an even coefficient from an odd one by multiplying by two.

2) The ***complex/simple*** rule: Place coefficients in front of more complex formulas before doing the same thing with simpler ones.

You'll quickly see how helpful these can be..........

In essence, once you can properly translate from "English" to "Chemical" language, writing and balancing equations are a matter of trial and error. Ultimately, I just tell my students to balance them and turn them loose to the task.

YOU MUST HAVE A BALANCED EQUATION BEFORE YOU CAN SOLVE ANY PROBLEMS!

YOU MUST HAVE A BALANCED EQUATION BEFORE YOU CAN SOLVE ANY PROBLEMS!

YOU MUST HAVE A BALANCED EQUATION BEFORE YOU CAN SOLVE ANY PROBLEMS!

Are you getting my drift?

STOICHIOMETRY (QUANTITIES): It's beginning to look a lot like chemistry!

Let's get some stuff straight first.

1. You're only going to use algebra (spelled A R I T H M E T I C), and rather simple algebra at that.

   Now some of you might be thinking, "I'll get by with cross-multiplication." You might, but why not use the factor label method as your instructor taught you? That's for your ease, not your burden. With that in mind, I'm going to show you a simple roadmap that you can use to solve quite a variety of problems.

Grams → Moles → Particles (atoms, ions, molecules, e.g.).

To convert grams to moles, divide the grams by the molar mass of the substance.

To convert moles (or parts of a mole) to particles, multiply those moles or parts by Avogadro's Number (no area code, 😆), 6.022×10^{23} – *my advice is to learn this number asap!*

To reverse the process, use the inverse operations; divide by Avogadro's Number and multiply by the molar mass, respectively.

Since you're reading this, your instructor is probably teaching you about grams, moles, and particles, and how to interconvert these quantities.

If you learn the map above, you're almost there; we're going to expand it just a bit very soon.

2. In dealing with a chemical process (reaction), you will invariably describe it by a **balanced equation**, as stated above. The substances on the left side of the arrow are known as the ***reactants***, and those on the right the ***products***.

Unless we're doing exploratory research, the real world typically requires solving for the amounts of reactants. That's right, get used to it! In industry, we know the amount of product we need, the "end result" if you will, and we need to solve for the amount of raw material we have to react to obtain it. This translates into cost to run the reaction, which leads to feasibility, etc. Are you getting the idea?

If we're designing clothing, we're going to have an end product in mind (dress, pants, coat, suit, etc.), the number of units of each that we need. What we'll need to calculate is the total amount of raw cloth we need to obtain to make all that product, and finally the cost of production. Likewise, pretty much all of my lecture class examples are practically-oriented.

Let "K" = the known substance, and let "U" = the unknown substance. Our roadmap becomes

Grams K → Moles K → Moles U → Grams U

<u>REMEMBER — BALANCE THAT EQUATION; OTHERWISE FORGET ABOUT SOLVING!</u>

I use **K** and **U** because, as I implied above, the **unknown** substance is quite often the **reactant** rather than the product.

Depending on whether I'm asked to find moles or grams, I plan a **route**: I get on the road and start from what I am given and arrive at my desired destination. Keep in mind that the mass (weight) of substance is more practical to find.

Now, pay particular attention to the arrow between moles K and moles U. This is crossing the Rubicon, so you better get this right!

24

Hey, Dr. C., what do you mean?

To cross the Rubicon, you will take the moles of K in the roadmap, and then multiply by the coefficient next to U **in the equation** and then divide by the coefficient next to K **in the equation.**

Let me illustrate by magnifying
that central section of the roadmap:

Grams K → Moles K → Moles U → Grams U

Let's blow up and analyze the blue region above:

$$\text{moles K (given)} \times \frac{\text{moles U (equation)}}{\text{moles K (equation)}} = \text{moles U (sought)}$$

You can easily see that the moles K divide out and we're left with moles U, after which we can continue to mass (i.e., grams) U.

Now do you see why the balanced equation is **essential**?

Try this example:

In industry, one of the ways nitrogen gas is produced is by the following reaction.

$N_2O_4(l) + 2N_2H_4(l) \rightarrow 3N_2(g) + 4H_2O(g)$

A customer calls and orders 30.0 kg of nitrogen gas. Calculate the necessary weight (mass) of N_2O_4 (blended with excess N_2H_4) to produce the desired weight of nitrogen gas ordered.

Here's the roadmap and route: kg N_2 → kmol N_2 → kmol N_2O_4 → kg N_2O_4

Kmol, kg??!! Are you crazy?!!

Oh, relax, you can convert to grams and moles if you need to, but why not just keep it all in kilovalues? (Did you notice – I could've responded equally in boldface, but the regular type makes me appear calmer and more collected. BTW, for your information, my electroshock therapist thinks I'm making EXCELLENT progress!)

If you followed the roadmap, starting with, let's say, 3.00 x 10^4 grams (30.0 kg) of nitrogen gas, and back-calculating, you should have come out with 3.28(4) x 10^4 grams (32.8 kg) of N_2O_4. So what is this number?

It's the absolute MINIMUM amount we require.
Now, to test your common sense thinking: Does that mean that as a chemist you are going to measure out 3.28(4) x 10^4 grams of to get what you (and the customer) wanted?

Of course not! I'M GOING TO ADD MORE THAN THAT TO BE SURE. THAT'S MORE REALISTIC, ISN'T IT?

Of course it is! You knew that intuitively. See? Left-brain/right brain! **Incomplete reaction, side reactions, loss of product in recovery,** etc. A whole host of problems that can't even yet be identified. Your answer might satisfy a textbook problem, but even food recipes get tweaked to compensate for inevitable loss or deficiency.

You see, the company has to deliver 30.0 kg of nitrogen gas to the customer by the end of the day. By the way, what the customer wants is the **ACTUAL YIELD** here, so we'd better come up with that. **The customer doesn't care how we get it.**

So how much extra raw material should we add? Well, as a guide, let's say that from company history and records, this process is known to have a 75% yield. We can use that statistic to find the answer. The first step is to use the schematic above. That result would assume 100% yield. Then multiply the result (see above) by the **improper fraction 100/75**. The new result is 4.38 x 10^4 g (43.8 kg) of N_2O_4. We would start with around this amount, which would help cut waste. I think you get the idea.

Let's get something straight *(wait - I think I said something like that earlier………..)*. **I AM NOT ASKING YOU TO DISREGARD THE IMPORTANCE OF FINDING % YIELD!** The fundamental equation there is AY (actual yield)/TY (theoretical yield), multiplied by 100. You will likely need that in lab. Furthermore, remember that in research and development, it is critical to find the % yield of a process so that middle or upper management can determine the financial practicality of incorporating it or going back to the drawing board to find a more efficient one!

THEORETICAL YIELD (as mass) refers to products; it's also referred to as the MAXIMUM POSSIBLE YIELD you could achieve. However, in the real world, we never get there. If somehow our lab balance told us we did get there or beyond, the sample was probably wet or otherwise contaminated. By the way, you can pre-calculate this theoretical yield before you actually do the experiment.

The **actual yield** (AY) is what you end up with after the experiment's done. Your lab instructor is likely to require a % yield, which again is simply (AY/TY) x 100.

To make matters slightly more complicated en route to calculating the theoretical yield, you may need to determine the **limiting reactant (LR)**, the one in short supply. It will therefore limit the "ceiling," or how much product you can make. Make sense? You have a recipe for cookies that requires two cups of flour and one stick of butter. You have the stick of butter but only 1.5 cup of flour, so you can still carry out the recipe but you'll have to adjust the butter quantity proportionately and you'll be limited in the number and/or size of cookies you can make.

Simple concept, right? Typically, instructors give problems involving at most two reactants and the weights of each. So, you MUST find out which is in short supply and use that to calculate the theoretical yield, etc., if asked.

This might shock you, but I don't elaborate about LR. I do show my students what appears to be the simplest method (see below), and then turn them loose. A caveat: if you want to use your own method, make sure you stick with the mole ratio principle. ***If you err in your quest for the LR, the rest of your calculations will be meaningless.***

Here's an example that requires finding the limiting reagent en route to the final desired answer:

For the process
2 SO_2 (g) + O_2 (g) $\rightarrow$ 2 SO_3(g)

14.2 grams of SO_2 are mixed with 18.5 grams of O_2. The reaction is run and 15.2 grams of SO_3 are **actually** collected.

a. What is the % yield of SO_3?

b. Middle management has stated that if at least 78% yield is not attained, it's back to the drawing board for the Research & Development team.

OK, so first **we'll pretend that SO_2 is the LR**, and we will calculate the "TY" based on that **assumption**. So, use the roadmap: g SO_2 → mol SO_2 → mol SO_3 → g SO_3. Do the calculations. Don't worry; I have all day (😄). The TY should be 17.8 g SO_3. Hold that thought.

Now, **pretend that O_2 is the LR**, and again calculate the "TY" **based on that assumption**. Same roadmap of course, but substitute accordingly. The TY now should be 92.5 g SO_3.

Which is the smaller number of the TY? That reflects the LR, which in this case is SO_2.

But you've killed two birds with one stone, if you know what I mean *(my deepest apologies to the animal rights groups)*. In the process, you calculated the TY (17.8 g SO_3). So now, what's left? Why, the % yield of course! Do the math and you'll arrive at 85.5%, so we can boldly tell management that we've satisfied their requirement!

Actually, every good chemist will repeat the process several times and take an average, so the results are more credible.

I know what you're thinking.

This stoichiometry stuff is overwhelming. How do I keep it all straight?

It may seem overwhelming, but here's a convenient summary to help you relax.

1. **Make absolutely certain that you have a balanced equation before you do anything else! (What? You mean you're tired of hearing that?)**

2. What is given? Masses, moles? Write them above each reactant or product. That's the "K" value. If you are given the weights of two reactants that are "mixed together," that's a giveaway that you are dealing with a **limiting reagent** problem. Use the above example to help you calculate the true theoretical yield (the "U" value here). Just use the roadmap and **draw your route**.

3. What does the problem ask for beyond any limiting reagent work? If it's percent yield, you ***MUST be given an ACTUAL YIELD*** in the problem. Then you know the formula: It's (AY/TY) x 100. Otherwise, whatever substance you know nothing or little about is the "U" value. Map out a route, and you're good to go.

You can either memorize all of the above or understand the roadmap. If you understand, you'll have an easier time tackling a rather wide variety of problems.

Aqueous Solutions:
Absolutely NO drowning allowed on my watch!

Aqueous solutions are simply homogenous mixtures that use water as the **solvent.** Speaking of **solvent,** you should know what that word means, as well as the meanings of **solute, solution, solubility, dilute,** and **concentrated. KNOW** what these are, and I recommend you demonstrate your understanding by being able to provide *examples*. Aqueous solutions typically involve a solid solute dissolved in water. However, keep in mind that they may involve gases or other liquids as solutes (e.g., soda water and alcoholic beverages).

By the way, know not only the terms above, but also **solvation** (**hydration** if we're referring to water). Solvation refers to the overall **mechanism** of dissolving the solute in the solvent. See below where I talk about solubility.

I won't waste time defining or describing **acids, bases, electrolytes** (strong/weak), or **non-electrolytes. Look these up and know them.**

is defined as moles of solute per liter of the resulting solution. Your instructor may include problems that ask you to calculate the molarity of a particular solution.

The basic roadmap for this is:
mass of the substance (K) → moles (K) of it → molarity

Obtain the molarity by **dividing moles K by the solution volume in liters.**

Here is how solution chemistry is grafted into our original roadmap:

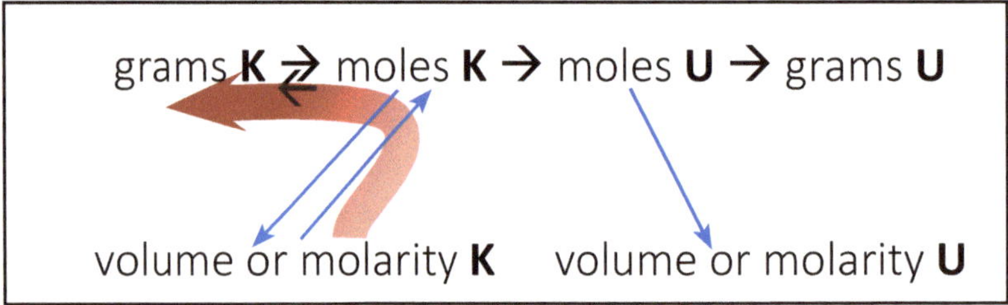

So why the double arrows in the K region?

A practical problem involves **solution preparation** (see a) below). You'll be starting from volume or molarity of K and finding grams of K to dissolve in the solvent.

So, what kinds of aqueous solution chemistry problems are you likely to encounter?

a. **Solution preparation** (If you're working in an academic research or industrial lab, you **MUST** know how to do this!)

b. **Solution dilution** (ditto)

c. **Stoichiometry** (just as in the earlier chapters with the simpler roadmap)

For a) and b) above, we'll stay on the left side (K) of the roadmap, as we're not dealing with a reaction. As was already mentioned, that's why I have double arrows on the K side. For c) problems, we'll need the full map.

Let's get to work solving a) and b) problems.

30

Your lab boss, after wishing you a good morning (I hope), tells you to prepare 2.00 liters of a 0.748-M solution of K_2SO_4. **Specifically** describe how you would accomplish that. By the way, what is the name of that solute above? **WHAT DO YOU MEAN YOU DON'T KNOW?!**

Step 1: (Now what do I do? Scratch head, pace the floor......**I hope not!) LOOK AT THE MAP, AND STAY ON THE LEFT (K) SIDE!!!!!** Sorry; I raised my voice just then!

1. Start with the map:

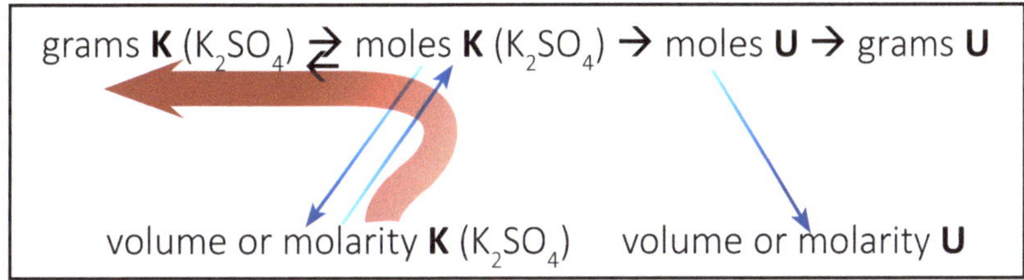

2. Plan the route: Follow the red arrow

Volume of (K_2SO_4) solution → moles of solute (K_2SO_4) → grams of solute (K_2SO_4). OK, ok......

3. Set up the math, preferably using factor label method. I'll get you started: Volume x molarity = moles. From the diagram you should be able to finish the rest of it.

    The mass of K should be around 261 grams.

4. Don't look at me! What do you think? That's right: Dilute the 261 grams to a total volume of 2.00 liters!

By the way, if you didn't mention under Step 1 that you would first obtain a 2.00 liter volumetric flask, **YOU FAIL! BIG TIME!** Securing the flask early is important because: 1) It reminds us that we're constrained to two liters, and 2) we're reminded of the precision of the volumetric flask (two decimal places).

Your opinion: Can you legislate molarity? (😄 couldn't resist it!)

Dilution

Dilution is part of **solution preparation**, and it is strictly a **one-way** street. The concept is taking a given # of moles and spreading them out in a larger volume of solvent. Imagine a can of frozen concentrated orange juice having several cans of water added to it. A certain brand of soup offers a concentrated version; just add water. You get the idea.

I know, I know: $M_1V_1 = M_2V_2$, but you still have to know what you're solving for. **IT DOES YOU NO GOOD TO SIMPLY SOLVE FOR "X" IN THIS TYPE OF PROBLEM.**

I like to say MV before the dilution = MV after the dilution. Since the number of moles doesn't change, that means that the molarity goes DOWN while the volume goes UP. Get it?

OK, let's do a problem.

That same boss later comes in *(probably after both of you have had at least two cups of coffee; btw, industrial coffee is typically spelled M U D)*, and tells you to dilute that original 0.748 M K_2SO_4 solution to 0.0546 M, and he needs 500 mL of it. NOW what do you do?

Why, the same process! But think...the molarity is going down, so I'll have to add water.

OK- Set up the problem the same way. That's right; (MV) before the dilution = (MV) after it.

We know that we need 0.0546 M solution and 500 mL when we're done ("after"). So the right side of the equation is taken care of. What about the left side? We know the original molarity (0.748 M). But there's a V term. What's that? That's the volume before the dilution. **IT'S THE VOLUME OF THE ORIGINAL SOLUTION YOU'LL NEED TO REMOVE AND DILUTE TO A TOTAL VOLUME OF 500 mL.** BTW, that comes out to 36.9 mL.

You will remove 36.9 mL of the original solution and dilute to a total volume of 500 mL.

But 36.9 is NOT a nice round number!

So, what's the problem? Stop whining! You may need a syringe or graduated cylinder for measuring, but you'll figure that out. In general, it's a good idea to ask your boss how precise that final concentration needs to be anyway, so you're not driven out of your mind.

(Too late; that ship has sailed! 😄 — Actually, in my case, it's sailed, foundered, and become the subject of a National Geographic documentary! 😄)

Here's another one. I'll give you the data, but you're on your own to solve it. I'll drop off the answer later.

We need 5.00 liters of a 0.475-M HCl solution for our experiments today, but the only solution available in the stockroom is 6.00-M. What do we do?

OK, set the problem up.

I'm waiting…Take your time; I ate last week!

OK, I got it: MV before = MV after.

Good start, and you know the "MV after," right? You also know the M before………hint… What's left?

If you solve for V (before), you should get around 0.40 (or 0.396) liters, or 400 (396) mL. What you will do is remove this volume of stock solution and dilute it to what? That's right, to a **total volume of 5.00 liters.**

Solution Stoichiometry Problems

These are the c) problems I mentioned above. Now we need the **full roadmap**, because we'll start with some type of "K" and travel to some type of "U." Ready?

Titrations

Titration is the simpler term for what's formally called volumetric analysis, and it's a rather important part of chemistry. You may be doing a titration experiment in lab this semester. The good news is that you can use our roadmap to solve this type of problem as well.

Example:

We have a field sample of HCl (**hydrochloric acid**, but you knew that, right?) of unknown molarity that we're trying to determine by titrating it with barium hydroxide. We pipet exactly 25.00 mL of the unknown HCl sample into an Erlenmeyer flask and titrate it with 0.846-M $Ba(OH)_2$. We find that the titration requires 28.23 mL of base to get to the visual endpoint. What is the molarity of the unknown sample?

What did I say you needed before solving? That's right; a balanced equation.

$$\overset{\textbf{K}}{Ba(OH)_2}\ (aq) + \overset{\textbf{U}}{2HCl}\ (aq) \rightarrow BaCl_2\ (aq) + 2H_2O\ (l)$$

Remember the modified roadmap?

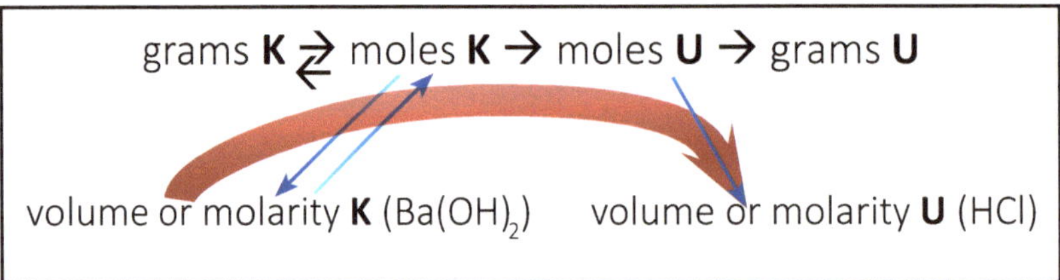

OK, so let's start with what we know most about: both the **molarity** and the **volume** of the **base** (barium hydroxide). Once you have the route, follow it (red arrow, above).

Actually, you should be adequately adept at the factor label method by this time to apply it to solution problems of this type.

Incidentally, your answer should be 1.91 M HCl.

You can solve virtually any volumetric analysis problem using that roadmap. Hopefully you're beginning to see a pattern here. We encounter the mole conversion each time, the core principle in stoichiometry!

We can use the above principles to solve **precipitation** problems such as the following.

Barium sulfate, due to its insolubility, has been used in radiologic studies to help diagnose diseases of the digestive tract. Calculate the minimum molarity of 3.00 liters of a potassium sulfate solution necessary to precipitate all of the barium ion in 2.00 liters of a 0.437-M barium nitrate solution.

Let's review: We're adding potassium sulfate (K_2SO_4) solution to barium nitrate ($Ba(NO_3)_2$ solution in order to precipitate barium sulfate ($BaSO_4$) solid, with potassium and nitrate ions hanging around (spectator ions).

1. Write the full balanced equation: What do we know most about (K)? The barium nitrate, right? Therefore, potassium sulfate will be the "U" substance.

   U   **K**

$$K_2SO_4(aq) + Ba(NO_3)_2(aq) \rightarrow BaSO_4(s) + 2KNO_3(aq)$$

BTW, you should be able to write the net ionic equation (NIE) if your instructor asks you. We'll keep the full equation for the problem, though.

2. Write the roadmap and the route.

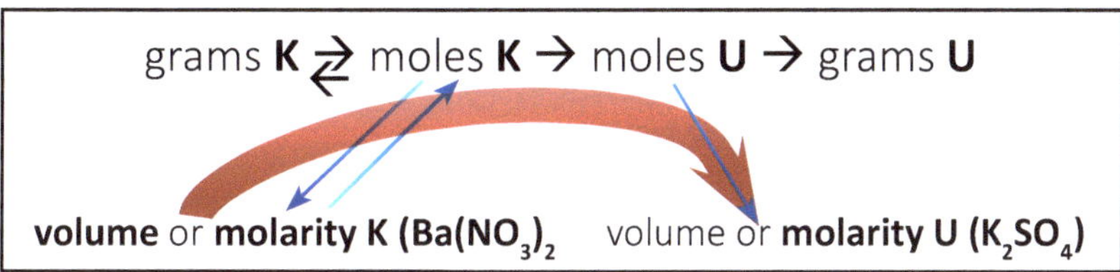

3. Solve. Your answer should be around 0.291 M potassium sulfate solution.

Here's another one, at least as practical as the one above.

Sodium bicarbonate (formula?) is a common component of antacids, which of course are ingested to neutralize hydrochloric acid (formula?) in an upset stomach. Calculate the volume of stomach acid that is 3.16×10^{-2}-M that can be neutralized by a tablet that contains 2.34 g of sodium bicarbonate.

I was needling you above to see if you remembered your formulas.

Anyway, what are we given in the problem? The mass of the sodium bicarbonate.

What do we need to find? The volume of the stomach acid.

The balanced equation will of course confirm we start with a "K" and end with a "U."

$$\text{K} \qquad\qquad \text{U}$$
$$NaHCO_3(aq) + HCl(aq) \rightarrow NaCl(aq) + H_2O(l) + CO_2(g)$$

Next, draw our map and the route:

grams **K (NaHCO$_3$)** $\rightleftarrows$ moles **K** $\rightarrow$ moles **U** $\rightarrow$ grams **U**

volume or molarity **K**   **volume** or molarity **U (HCl)**

If you understand the map and the route, you are 99% there! BTW, the answer is around 881 mLs of acid. Did you get stuck between moles U and liters U? Divide moles U by the molarity of U.

Solubility

Ready for another outrageous idea?

Everything is soluble (in everything)!

Are you crazy Dr. C.?! I was taught certain chemicals are not soluble in each other.

In context, it's hard to argue with you. But you will see eventually that that statement is not such an extravagant assertion.

I teach my students to learn solubility as a property of ***degree*** rather than of ***kind***. Generally, and particularly in this case, I discourage compartmentalized thinking from my students. Of course, you must learn the textbook solubility rules **as directed by your instructor.** But understand the basis of solubility.

Three forces are involved in the **solvation** (dissolution) process: **solute-solute, solvent-solvent, and solute-solvent** interactions. The degree of solvation (**hydration**, if water is the solvent) depends on the ***net*** result of these forces "battling it out."

36

As far as **net ionic equations** are concerned, these just take practice to write. The main problem I've found over the years is that students either haven't sufficiently learned solubilities or they don't yet recognize polyatomic ions within compounds. See below for some more detail.

Types of Reactions

Don't let this section get you upset. Reactions (described by equations) are classified various ways. I will attempt to streamline and simplify this for you. The appropriate appendix at the end of this book may also be helpful.

Lately, textbooks have been using terms such as precipitation, acid-base, gas evolution, and oxidation-reduction (AKA **redox**) as classification terms. Why is classification so important? Mainly, recognizing a reaction by its type can help you determine probable products of the process (say that three times slowly!). For example, finish this sentence: Acid mixed with base gives _____________ and _______________.

As with anything else, the more exposure to and experience with these reactions, the easier it will be to recognize their types. That's why studying daily is important.

For convenience, you will notice that precipitation, acid-base, and gas evolution reactions all appear to behave in essentially the same way: they exchange ionic components during the process to make new substances, kind of like two dance couples exchanging partners.

For example, in that acid-base problem I gave you above, your answer should have been salt (an ionic compound) and water. When HCl reacts with NaOH, by switching ionic components during the process, we get HOH and NaCl.

Now, in the above reactions, the products typically are a relatively insoluble substance (precipitate), water (stable molecular compound), or a gas (which escapes by bubbling out of the solution as it's formed). These collectively are what we call "driving forces" of reactions. They are stable substances that under the circumstances are not likely to react any further at the moment. ***Keep this principle in mind!***

By the way, all three of the above reaction types fall under the category "double replacement," which is just another way of stating "switching partners," as I mentioned above. Look up "double replacement" reactions when you have some time.

By now your instructor has probably told you that you need to be able to write out the *net* process (net ionic equation, see below example) that occurs with all three of these reaction types. We're back to that first **outrageous statement** I made at the beginning of the book: Why did the reaction occur in the first place? Answer here:

____

________________________________.

Good, you understand. The natural tendency is toward stability.

In the simple case above, the *original equation* (your textbook calls this the "molecular equation," a term I'm not comfortable with) was HCl (aq) + NaOH (aq) → H_2O (l) + NaCl (aq) (*O.E.*).

General rule: When you see the subscript "aq" for a salt, you can separate the substance into ions, but if not, keep it together!

So we could re-write the above equation as:
H^+ + Cl^- + Na^+ + OH^- → H_2O (l) + Na^+ + Cl^- .
This is known as the *total ionic equation, T.I.E.* By the way, you'll notice I left out the "aq"s to avoid the clutter in this step.

From here on in, it's a matter of math: Cross out the common terms just as you would in algebra with x's and y's. What are you left with? H^+ (aq) + ^-OH (aq) → H_2O (l). This is called the *net ionic equation, N.I.E.*

The ions we crossed out are called *spectator ions*, a fitting term as they really don't participate in the reaction, do they!

You can do the same three-step process with precipitation reactions and with gas evolution reactions! Just follow the rules.

Oxidation-reduction Reactions

You're going to get an earful of this in your lecture class, but the short story on oxidation-reduction (redox) processes is that they reflect electron transfer within the reaction. That's it! Oh, you have to learn about oxidation numbers, which are collectively just a bookkeeping device. Many of these numbers you already know as numerically equivalent to the ionic charges.

Redox reactions can incorporate different types of compounds. This requires some detective work on your part. For example, you will commonly be asked whether a certain reaction is a redox one. Here's how you can tell: **LOOK FOR A CHANGE IN OXIDATION NUMBER!** That's it. For example, in the process

$$2H_2\,(g) + O_2\,(g) \rightarrow 2H_2O\,(g)$$

Right away you can see that the oxidation number of hydrogen starts at 0 at the left of the arrow and goes to +1 to the right. We could continue and show that oxygen's oxidation number simultaneously changes from 0 to -2 during the process.

I have found that using a number line is helpful as a visual aid in teaching and learning the concept of redox.

Oxidation

-6 -5 -4 -3 -2 -1 0 +1 +2 +3 +4 +5 +6

Reduction

The oxidation direction is **less negative to more positive**, and vice versa for reduction. Get the idea?

So with redox reactions, you only need to identify/verify the elements whose oxidation numbers change during the process. As I tell my students, it's like the pitcher-catcher relationship: If somebody is pitching the ball, someone else is there to catch it (Law of Conservation).

***Your instructor will probably want you
to know all the redox rules!***

I won't elaborate, partly because you already have a chart in your textbook. Here's the good news: When you need to calculate any missing oxidation numbers, the math is very easy.

Gases

Gases are easy to study because over a fairly wide range of temperatures and pressures, their observed behavior can be described by very **simple equations**. No doubt your instructor is going or has gone over the basics of gases and a discussion of the nature of **pressure** (force/area).

So I'll cut to the chase. All those gas laws converge to **PV = nRT**, the **Ideal Gas Law**. Within this expression you can see all the other laws that evolve to it.

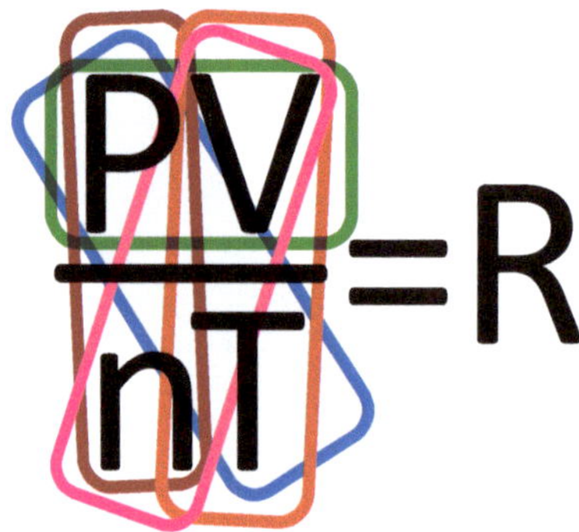

Now, we will dissect that equation, so make note of the following relationships: PV, V/T, P/T, V/n, and P/n. Take a moment to examine the Ideal Gas Law equation above. Do you see these five circles within it?

Now, I don't believe that it's important that you know that a certain law is Boyle's, Charles', etc., **although your instructor may**. What _is_ important is that you understand that:

1) The evolution of the laws involves four variables, namely, **P**, **V**, **T**, and **n** (the number of moles of gas), and

2) In each of the laws, there's an independent variable ("x") and a dependent one ("y"). Each time **all other variables are stipulated to be constant**. There's nothing magical or exotic about this idea. When scientists do lab experiments, they

typically keep all conditions constant except for one, and see how any other (dependent) variable changes when they change that one.

For example, the **PV** relationship involves keeping the **T** and **n** variables constant, and then we'd like to see how the volume (V) varies with changes in pressure (P).

Now, so that we keep the left brain/right brain dynamic, **VISUALIZE** two models: the **balloon/piston** model and the **rigid gas cylinder** model. The first will be useful for the laws where volume changes are involved (PV, V/T, V/n), and the second will be useful where there are no volume changes (P/T, P/n).

Fundamental Law	Stipulations	Operative Form
PV = Constant (Boyle's)	T, n (# moles) are constant	P1V1 = P2V2
V/T = Constant (Charles')	P, n are constant	V1/T1 = V2/T2
P/T = Constant (Gay-Lussac's)	V, n are constant	P1/T1 = P2/T2
V/n = Constant (Avogadro's)	P, T are constant	V1/n1 = V2/n2
P/n = Constant (Dalton's)	V, T are constant	P1/n1 = P2/n2
PV/nT = Constant (Ideal)	None	(P1xV1)/(n1xT1) = (P2xV2/(n2xT2)

Look at the last line in the table above. That's right: **PV/nT** is a constant. All we did was mash, to use the vernacular (😄), all the laws together, which is to say, all the variables together.

You know, if you analyze these laws using one of the two models I described earlier, gas behavior is intuitive.

- **PV**; the volume on the gas of the balloon shrinks as the pressure on it rises. Makes sense........

- **V/T**; the volume of the gas in the balloon expands if the temperature is raised, as the molecules are moving more energetically. Makes sense..........

- **P/T**; the pressure within the rigid container rises as the temperature does, using the same reasoning as with Charles' Law. Makes sense....

- **V/n**; adding more molecules to the balloon raises the volume. Makes sense.......

- **P/n**; adding more molecules to a rigid container causes more collisions and raises the pressure. Makes sense....

A practical note...Look at P/T. This is the reason you NEVER heat a closed system (e.g., a stoppered flask filled with a liquid chemical). An explosion can occur rather easily, or an overfilled gas cylinder could become a missile. This has actually happened! Back in the '60s (I know, ancient history), soda was sold in glass bottles. Occasionally, due to warm environments, the pressure of the CO_2 inside the bottle got too high, and the bottles exploded. I had heard that several shoppers suffered eye damage because they were struck by flying glass.

Now, tell me, what are you NEVER going to do?

R - The mystery constant

Now, how do we arrive at R? Once upon a time, "they" (certain scientists) figured out that one mole of ANY ideal gas (pick a gas, any gas....) occupied 22.4 liters at 0°C (273 K) and one atmosphere (760 mm Hg) of pressure. Now, y'all, plug those numbers into the **PV/nT** expression, and what do you get???? Why, **0.0821 L-atm/mol-K**, that's what!

Now, re-do the calculation, but this time use **760 mm** for the pressure, and you should come up with **62.4 L-mm Hg/mol-K**.

So, which answer is correct?

BOTH! That's right; *it depends on the units of pressure used in solving the problem*. Just remember that. By the way, that answer is, per the table above, a CONSTANT, called **"R," the universal gas constant.**

*As a post-script, that **22.4 L/mol** is useful **ONLY** under the conditions of 0°C (273 K) and 1 atmosphere of pressure, conditions not normally encountered in the general chem lab.*

Applications

Now that you know the gas laws, here are some uses:

- Density determination

- Molar mass determination

- Stoichiometric problems (using V/n); you'll see how we graft in the volume-mole conversions into our chart

- Partial pressure determination in a gas mixture (P/n)

Your instructor will likely derive these equations for you. It's probably most expedient if you memorize them. However, do watch what your instructor is doing; density and molar mass expressions (see below) are derived from **PV = nRT.**

The density of a gas in #grams/liter = (P x MM)/RT

Look at the density equation and analyze it. **Pressure** and **molar mass** are in the **numerator**. Makes sense; higher pressure brings the molecules together, and of course higher mass naturally increases the density. On the other hand, raising the temperature (denominator) raises the average kinetic energy and tends to spread the molecules out, lowering the density.

The molar mass of an unknown gas (grams/mole) = (#grams of gas sample x RT)/PV

We won't agonize over the second one. But if you do a **unit analysis** of each of these equations, you will end up with **g/L** and **g/mol**, respectively.

Since the above types of problems are essentially plug-in, I won't waste time with examples. Just watch the numbers and carefully cancel out units

Remember, when you're doing calculations:

- **The temperature MUST be in Kelvin to solve gas problems.**

- **MAKE SURE you have the right R value depending on the units of pressure P. R is mainly either 0.0821 L-atm/mol-K, OR 62.4 L-mm Hg/mol-K.**

We can solve other stoichiometric problems by modifying our roadmap with an additional entrance and exit, just as we did for solution chemistry.

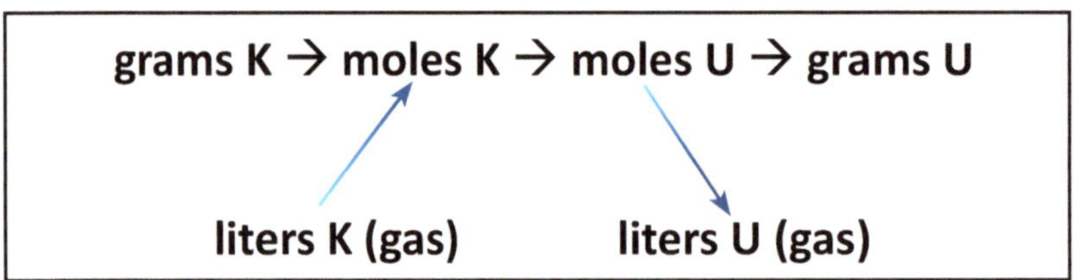

Here we apply Avogadro's Law (V/n), which says that in an equation, the coefficient in front of any gaseous substance can be thought of as either the volume (in liters, for example), or as the number of moles of gas.

Let's illustrate: $2 H_2 (g) + O_2 (g) \rightarrow 2H_2O (g)$.

I can read that equation two ways: "Two moles of hydrogen gas react with one mole of oxygen gas to produce two moles of water (vapor)," or "Two liters of hydrogen gas react with one liter of oxygen gas to produce two liters of water vapor."

To refresh, here's the operative form: V1/n1 = V2/n2, a **direct** relationship. But here's the stipulation:

THE TEMPERATURE AND PRESSURE ARE FIXED!

So how do we use the gas law in this type of a problem? Gases are often seen in chemical equations, and we may want to find the volume of a gas under certain reaction conditions.

Example:

For the reaction, $NH_3 (g) + HCl (g) \rightarrow NH_4Cl (s)$, what volume of ammonia gas will we need in excess HCl gas to produce 12.6 grams of ammonium chloride, if the reaction is conducted at 26°C and 0.879 atm pressure?

OK, follow the roadmap , applying the Avogadro law along the way. What is our K? 12.6 grams of ammonium chloride of course. What is our U? Look at the map above and identify it. It's liters of ammonia gas. Just plan the route and follow it, as illustrated below:

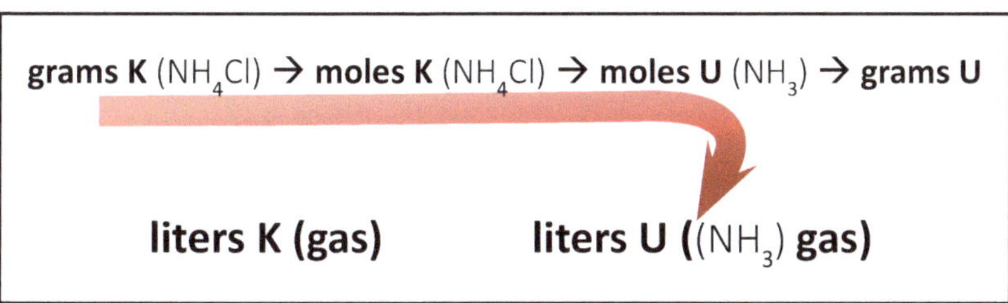

OK, Dr. C., I'm fine until I have to go from moles of ammonia to liters of ammonia.

No problem. Just use **PV = nRT**, only manipulate to get **V = nRT/P**. By the way, which value of R are you going to use?

Oh yes, the answer to the problem should be **6.57 liters of ammonia gas**.

Also, don't be frightened if you have to solve **limiting reagent problems** involving gases. They're very similar to the ones you've already solved. You may just need to **convert liters to moles using the Ideal Gas Law** as we did above.

Applications of Dalton's Law: "They're all the same."

No doubt your text and your instructor have pointed out Dalton's Law of Partial Pressures. Pay attention to this for several reasons.

- If you have a gas mixture, you can figure out the pressure contributed by each gas if you know the amount (moles) of each gas present. Remember, the **pressure is directly dependent on the number of moles of gas** we're talking about. P1/n1 = P2/n2.

- Another piece of good news: you simply add up the partial pressures of the component gases in the mixture to get the total pressure.

- If a gas is collected over water, which often happens in solution chemistry where the reaction produces one or more gases, we must subtract out the vapor pressure of water at the particular temperature of the lab that day (look it up – you can Google it). This will give us the pressure of the dry gas alone. P_T (the total pressure; Google it or read the lab barometer) = P (gas collected) + P (water vapor). The pressure of the water vapor will depend on the temperature, so have a thermometer handy.

Therefore, P (dry gas alone) = $P_T - P_{water\ vapor}$

I won't overburden you with relationships. Just remember this one:

$P_{total} \times X_{gas\ A} = P_{gas\ A}$ (X is what's called the mole fraction of that particular gas compared to the total number of moles of all gases in the mixture. It's a **unitless** quantity because it's a fraction.

Don't worry; I'll translate. Since the pressure contributed by any single gas component in a gas mixture is proportional to the number of its molecules there (P/n, remember?), if I multiply the **total pressure** (P_{total} above) by the fraction of the molecules which are that single gas, I should get a number which is the actual pressure contributed by those particular molecules. Make sense?

Root Mean Square Speed and Graham's Law

You'd better know these equations, just in case your instructor decides to test you on them. What are the essential principles? **Gases move faster at higher temperatures, and heavier gases move more slowly**. Intuitive, isn't it?

The KMT (Kinetic Molecular Theory)

The gas chapter includes both **law** and **theory**. Quickly now, what's the difference? Here are the postulates of the KMT.

- Gas particles move in straight lines and collide frequently and elastically (look it up) with their neighbors.

- Gas particles exhibit no particular attraction to or repulsion from their neighbors. (This is why we can mix gases and they behave as if they were all one gas.)

- The volume of each gas particle is negligible compared with the total gas volume. My chemistry professor way back when used to call these individual volumes "dimensionless points in space." If you think about it, it makes sense since gases are rather spread out, aren't they?

- The **average kinetic energy** of a gas directly depends on the Kelvin temperature. This should make immediate sense to you. We are dealing with molecular motion. By the way, always keep in mind that we are dealing with billions upon billions of molecules, not just one or two, so the term **average** should conjure up a sort of bell curve in your mind.

Be able to casually discuss these postulates **IN YOUR OWN WORDS**, and not necessarily in any particular order.

The van der Waals equation ("Real Gases")

When pressures are sufficiently high **AND** temperatures sufficiently low, gas molecules get rather close to each other. As a result, their **individual volumes** and their **intermolecular attractive forces** become significant.

So, the van der Waals equation represents an adjustment of the simple PV = nRT expression to account for these changes.

When I lecture, I typically don't have students memorize this or work with it beyond discussing the main point. However, your instructor may require you to do some math with it!

Thermochemistry

Just as we have the **Law of Matter Conservation**, we have a parallel one for energy. I will let you state that rather than writing it out here.

Our energy source on earth is ultimately the sun, so we can say that solar power is the primary type of energy. It then gets converted either directly or indirectly to one or more forms, including **heat**, various types of **electromagnetic** ("light"), **chemical** (our famous chemical bonds), and of course **mechanical** (the famous "work"). You could get even more specific beyond these as I'm sure you're aware. Of course, let's not forget classifying energy as either potential (position), or kinetic (motion). Since molecules always exhibit some type and degree of motion, from a practical standpoint, that's about as concise a definition of energy as you can get. ***Energy (specifically here the heat energy) = motion!***

OK, motion of WHAT?

Motion of molecules! As long as there's some kind of molecular motion, we have heat. Now, that heat may be of a vanishingly low level, but you get the idea.

The actual heat within a body or object cannot practically be measured itself, so when we talk about heat energy, we really mean heat **transfer** (ΔH, or sometimes **q**). We can also call it heat **exchange**, heat **change**, or heat **flow**. File these terms away; they're all synonyms.

Let's get organized. In essence, here's what you need to learn.

1) In chemistry, when something happens, it's called a **process**, as mentioned above. It might be **physical** (i.e., ice melting) or **chemical** (a reaction). Either way, it's called a **process**. We mention them here because heat is **transferred** regardless of whether the process is physical or chemical.

2) The **process** we're interested in, and **ONLY** that process, is called the **system**. Literally EVERYTHING else is called the **surroundings**. For example, the other chemicals (water, if a reaction is carried out in it), the beaker, the lab bench, the lab, and so on, are part of the surroundings.

3) Heat **ALWAYS** flows from a hotter area to a colder one, not vice versa!

When you feel cold during, say, the wintertime, heat has flowed from your body to the outside environment. In the summer, you feel hot because ________________________________ (oh, come on, you can fill that in!)

By the way, when heat flows away from any reference point (**system**), your body for example, that's called an **exothermic** process. Exothermic is Greek for "heat going out." By the way, what do you think **endothermic** means? (😁). In everyday lab chemistry, we encounter more exothermic processes than endothermic ones.

So, what's the big deal?

You will come across the terms **"q"** and **"ΔH,"** as I mentioned above. They will either be negative (-) or positive (+). Either term refers to the **direction** of heat flow or transfer. The negative sign means heat flows **out of the system** (**exothermic**) to the surroundings and vice versa.

Remember this fundamental point: In order to have a chemical reaction, you must break chemical bonds and reform them in some other manner to make new substances. That's where the chemical energy is stored, and as I said earlier, the overall effect typically (not always) is a net release (-) of heat energy.

Work

Sorry to complicate your life, but your text says there's another form of energy we must consider besides heat in this chapter: **work (w).** Do you see in your text where it says that the total **internal energy E** of a system is made up of **heat (q)** AND **work (w)**? **STOP READING AND LOOK THESE UP <u>NOW</u>!**

Again, my apologies, because after laboring to explain the work function to you, it turns out that work typically contributes negligibly to the total internal energy we calculate. Frankly, the only time in a chemistry lab we'd be interested in the work function is when we're dealing with expansion and contraction of large volumes of gases. To be safe, memorize the energy equation and the signs of the work function if your instructor requires.

A further caveat: Your instructor will use the term "PV" and "w" interchangeably for work, as he/she will do with "ΔH" and "q" for heat. See below (State Functions) for the explanations.

I tend to minimize the lecture time spent on the work function, because I don't think it has much practical significance for my students, particularly in the lab. But, of course you MUST follow your instructor's directives as far as that's concerned!

State Functions

If you don't completely understand this concept right away, don't sweat it (😄). In a nutshell, we use capital letters (e.g., **E, V, P, H**) for state functions.

The capital of New Jersey is Trenton. Sorry, couldn't resist it! 😄

OK, it just means we care about only the **initial** and **final** states during a process, such as a reaction, not the path the system took to get there. It's like you and I arranging to meet at a restaurant at the Jersey shore, beginning at a certain street corner in Philadelphia. You will take one route and I will take another. But we'll both end up at the shore destination (*At my age, I'll probably make a rest stop along the way*.) Another way of stating this is that only the **start** and **end** points matter, not the route.

The enthalpy function (H) is derived from the total internal energy E and the work function (w, or PV), long story short. I'll let you study that derivation in your text. Again, I don't want to reinvent the wheel. Suffice to say that practically speaking, **ΔH** is the old "q" made into a legitimate state function, as you will see in lab.

I'm trying to steer you back to the heat concept again. I stated this above but I'll restate it to assure you: The good news is that in the gen chem laboratory, the work function is usually if not always so small that E is approximately if not exactly equal to H. You see, heat, not work, is typically the lion's share of the energy transfer during a process.

Calorimetry: Measuring the Heat Exchange

Calorimetry is merely the Latin term for heat measurement. There are basically two types, **constant pressure** calorimetry and **constant volume** calorimetry.

Constant pressure calorimetry uses the typical equipment such as beakers, flasks, etc., in a common laboratory setting, open to the atmosphere. The "coffee-cup" calorimeters (look it up) are commonly used here.

Constant volume calorimetry has rather more sophisticated and precise instrumentation, as it is used to obtain the calorie content of food and fuel. This utilizes a **bomb calorimeter (look it up; it's not what you might think),** within which a sample is electrically ignited in an oxygen rich environment (to help it burn of course) in a fixed volume.

I need to discuss **temperature** with you and its relationship to **heat** before we get into problems. Be patient.

Don't you split hairs with me! I've about had my fill of thermochem already!

Hold on, hold on, we're almost there.

Heat is NOT merely **temperature**. Temperature is only one facet of heat, namely its **intensity**. For example, you read the thermometer outside the kitchen window to get the air temperature, or the **intensity** of the heat in the air, but that's as far as it goes.

There are two other components of heat. The first of these is the **sample size** (weight), because all those molecules are going to be involved in the heat transfer. The second of these is an internal (you learned it as **intensive**) factor, namely the **specific heat**, aka **specific heat capacity**, or in the case of constant volume calorimetry, simply the **heat capacity**. All these terms may seem insufferable to learn, but trust me, you should be able to learn them with the use of examples.

The **specific heat/specific heat capacity/heat capacity** terms refer to the internal ability of the substance in question to transfer heat. It is **specific** for that substance alone. The number you see when you look it up indicates how much heat must be injected into it or removed from it (typically one gram) to observe a temperature change of one degree Celsius. In the case of **constant volume** calorimetry, we're considering the heat capacity of the calorimeter only, which is specific for each calorimeter. The units there are typically kJ/°C.

In calorimetry in general, remember this principle, an application of the **Heat Conservation** law:

Heat out = Heat in

Let's translate: The heat given off or taken in by the system is absorbed (or given off) by the surroundings. Intuitive, right? Of course! By the way, y'all, as I mentioned earlier, quite often water is part of those surroundings.

Anyway, the operational form is,

$$-q \text{ (out)} = +q \text{ (in)}.$$

Since, I presume, you've been introduced to **state functions**, if we define the conditions, we can also write,

$$-\Delta H \text{ (out)} = +\Delta H \text{ (in)}$$

We'll get to the expanded form of the relevant equation in a minute. First, let me emphasize that you **MUST include a negative** (-) sign on any side. It doesn't matter which side, but it MUST be there for all the math to work out. **DO WE UNDERSTAND**? It's critical.

OK, so let's translate a bit more. In the **constant pressure** scenario, "q" (the heat transferred) = total mass of the sample (grams, typically) x specific heat or specific heat capacity (J/gram-°C) x Δ T (°C).

For the **constant volume** scenario, q = C (heat capacity term, J/°C) x delta T (°C). By the way, the physical translation of that "C" term refers to the characteristics of the metal calorimeter actually used in this type of experiments. C is a constant that is unique for every calorimeter and MUST be predetermined by experiment.

Congratulations! You've paid your dues. Now here's the reward: After enduring all the above, you will see how ridiculously easy the math is!

Here are some problems.

Example 1

Calculate the heat lost when a 75.9 gram sample of gold is cooled from 88°C to 26°C.

ΔH = mass x sp. ht. x ΔT = 75.9 g x 0.129 J/g-°C x (26 − 88)°C.

Hey Chesloff, why did you subtract 88 from 26?

Because in heat problems, ΔT is **ALWAYS** the **final minus the initial** temperature ($T_f - T_i$). Yes, it gives a negative result here, but remember that heat is being lost or given off **BY THE SYSTEM**!

Your answer should be -609 J.

Example 2

A piece of iron weighing 764 grams at 154°C is plunged into 1500 grams of water at 28°C. By calculation, predict the final temperature of the system.

Quick – visualize **cold** milk being added to **hot** coffee. Roughly, where will the final temperature of the mix be?

Well, here you use the same equation……..twice! Somebody is losing heat, and somebody is gaining it. If we assume zero net heat loss, we can say that the **heat lost by the iron is picked up by the water**, or – ΔH (iron) = + ΔH (water). Make sense? Heat transfer. OK, then all we need to do is substitute as follows.

-(764 g iron x 0.450 J/g(iron)-° C x (X- 154)°C) = +(1500 g water x 4.18 J/g(water)- ° C x (X – 28 °C).

DON'T FORGET TO INSERT the negative sign! That's right, either side, but it MUST be there!

Do you understand the set-up? The "X" must be the same value, because it's the **final temperature of the system** after the mixing, just like the milk in the coffee. Realistically, it will be a value **intermediate** between the two initial temperatures, not exactly between them.

Actually, you knew that intuitively, didn't you?

If you do the arithmetic (OK, algebra), X turns out to be around **34.5°C**, which is of course **intermediate, but not exactly,** between 154°C and 28°C, as we would intuitively expect.

Wait a minute! Why is the answer so slanted toward the initial temperature of the water?

Good question. Look at the specific heats. Water's is so much higher. It takes more "oomph," (translation: **heat energy**) to see a change in water's temperature than for iron, so I would expect that on average, the final temperature of this mix is going to be closer to the initial temperature of water than to that of iron.

Example 3

Here's one where we use a bomb calorimeter, and you'll see both the ease and utility of this calculation.

A certain hydrocarbon fuel is combusted in a bomb calorimeter by electrically igniting 1.745 g of it in oxygen gas. The calorimeter is immersed in a water bath (look in your textbook for a picture, or Google for one). The temperature of the water bath rises from 23.86°C to 26.77°C. If the calorimeter has a heat capacity of 12.8 kJ/°C, calculate the heat per gram of hydrocarbon.

OK, y'all, the heat sent out by the burning hydrocarbon is taken up by the calorimeter, essentially (OK, ok, the water bath).

So, **q (calorimeter) = -q (process, in this case the combustion)**. The math is:

q (process) = -(12.8 kJ/°C x (26.77 − 23.86)°C) = -37.2 kJ.

Hold on, we're not quite done yet. That was for the sample weight of 1.745 gram. To get the heat per gram, simply take the heat for the process and divide by the sample weight (-37.2 kJ/1.745 gram) to get **-21.3 kJ/gram** of hydrocarbon.

I hope you can see from the above the irony in all this; beyond all the complex jargon about energy/heat changes, the calculations are really simple and straightforward! No joke, really.

A quick post-script: Don't go too crazy over the signs. Yes, they're important, but most of the time if not always in chemistry they signify merely a **direction** within a process.

Heat Stoichiometry

OK, before we leave the measurements behind, I'll need to add an example about the standard heat produced (or absorbed) when a reaction occurs. So, in that case, what do we need before we do any calculating? That's right, a **balanced equation.**

For example, for the decomposition of mercury (II) oxide,

$2HgO(s) \rightarrow 2Hg(l) + O_2(g)$, **ΔH** = +181.6 kJ as written.

What does this mean?

When two moles of the mercury oxide break down, when two moles of liquid mercury are formed, and when one mole of oxygen gas is formed, the system **ABSORBS** (+ sign) 181.6 kJ of heat. Got it? As we prepare to do a real world problem, remember what I said way back; **treat energy like matter: quantitate!**

We need to recover 23.8 grams of liquid mercury from the above process. How much heat will we need to pump into the system to accomplish this? Let's inject some practicality. A company may need this information beforehand, because that translates to energy cost. Is it worth it?

Here's where energy fits into our roadmap.

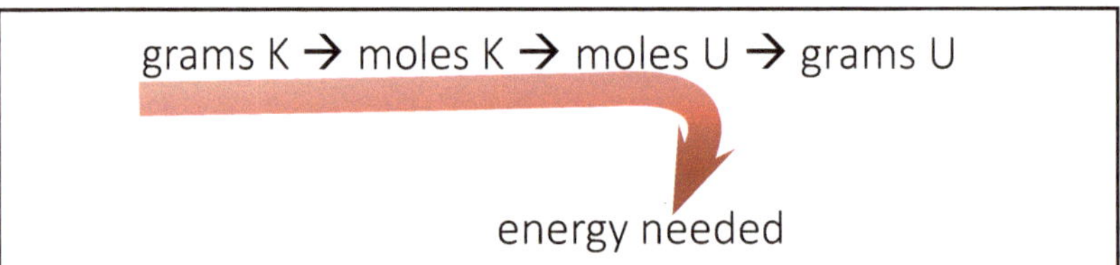

The energy term since energy is neutral for the process, So we don't need to designate either "K" or "U" here. We will use K for the liquid Hg, however.

BTW, the answer should be **+10.8 kJ**.

As you can see, the calculation is fairly simple after all! By the way, this should make sense intuitively also!

23.8 grams Hg x (1 mole Hg/200.6 g Hg) x (+181.6 kJ/2 moles Hg) = kJ total energy needed.

As you can see, the balanced equation is essential! **MAKE SURE** you use the correct coefficient in the calculation; not 1, but 2 moles Hg.

Standard States

ΔH° signifies the heat exchange under **standard conditions**. Don't get upset over this. Again, we need reference points for our calculations. The **ΔH°** of any process is ultimately derived from something called **ΔH°** of formation, which is arbitrarily the heat (+ or-) exchange that occurs when one mole of the substance is formed from its elements at 25°C. There are some other details. Look in your textbook, and I am sure your instructor will be fairly thorough about the stipulations.

Three items about dealing with thermochemical equations:

a. **If you reverse the reaction, you change the sign of ΔH.**

   Does it **INTUITIVELY** make sense that an exothermic reaction will require that same amount of heat to be put into it if it goes the other way?

b. **If you double all the coefficients, you double the amount of heat.**

- Does it **INTUITIVELY** make sense that if you double the quantities you combust, for example, you double the heat output?

c. **State functions imply that we can write a process as one step or as the net sum of multiple steps.**

   That's stated in your text as **Hess's Law**. Look it up. I'll let your instructor elaborate, but it turns out to simply be another math puzzle, kind of like one of those daily puzzles in the newspaper. You manipulate and then add a bunch of equations together and cancel out the terms on either side to end up with the net equation you want.

Quantum Theory/Electron Configuration

By now your course trajectory is arcing rapidly toward electron configuration in atoms and eventually chemical bonding. Remember, electrons are the subatomic particles that are actually involved in the bonding process. Keep in mind, however, that it's more complicated than that. Right now, we'll just take a look at electrons.

When I teach, I break this chapter down into two sections: the *matter-wave* section, and the *electron configuration* section.

To begin, you need to know about waves. Very simply, these are the media over which energy travels. The electromagnetic waves that travel through space allow your radio/television/cellphone to function. Waves obviously carry critical messages we need to function.

There are three components of waves you need to know:

- **-length** (distance between crests or troughs; units are *meters* or *nanometers*),

- **-frequency** (waves per second; units are s^{-1}, aka *cycles per second*, aka *Hertz*),

- **-amplitude** (height or intensity of the wave; units are *meters*).

The product of wavelength and frequency is a constant, namely the **speed of light (c), 3.00 x 10^8 m/s.** So, as the wavelength goes up, the frequency goes _________________(?) Vice versa, of course. Make sure you can interconvert wavelength and frequency.

- **Frequency magnitudes are typically on the order of 10^{14} s^{-1}.**

- **Wavelengths typically range from about 10^{-9} to 10^{-7} m. Your instructor will probably express these in nanometers (nm) because those numbers are easier on the eyes.**

I mention this because sometimes students make calculator mistakes by not hitting the right buttons and coming up with outlandish magnitudes of numbers. BE CAREFUL! What a shame to lose points because despite setting up the problem correctly you reported the wrong answer. **CHECK YOUR ANSWERS** always! Are they **reasonable?**

Finally, the "speed of light" is a rather loose term that really refers to the speed of electromagnetic radiation (EMR), as it pertains to the entire electromagnetic spectrum (EMS) your instructor will no doubt reference. You will be using the same constant regardless of whether you are calculating in the TV/cellphone region or the x-ray/gamma ray region.

OK, now you will need to know about some scientists who contributed to quantum theory development. I'll give you the capsule versions as always.

Planck (~1900) – Articulated quantum theory. Up to that point, scientists thought energy was absorbed or given off continuously, but a lot of disconcerting data was noted about this. Planck asserted that energy is absorbed or given off by matter in packets, or quanta, as he called them.

- **One quantum of energy = Planck's constant x frequency, or h x v(nu, Greek letter). Planck's constant (h) = 6.63. x 10^{-34} J-s.**

- **Energy quanta should be on the order of 10^{-18} – 10^{-20} J each. Again, your instructor will probably want you to adjust this to kJ/mol, only because one quantum of energy simply isn't of any practical use to anyone. Just convert the energy term (J) to kJ, and then multiply by Avogadro's Number.**

By the way, you should **LOVE** the quantum approach; **IT'S DIGITAL!** It fits right in with your life. Let's face it; the entire world now runs digitally.

Einstein (early 1900s) – Invoked the quantum theory to explain the **photoelectric effect** (look it up). Critical: Einstein proposed the idea of light **PARTICLES** versus **WAVES**; hence the particulate nature of light, or **_photon_**. Be able to explain the photoelectric effect in your own words!

BTW, Einstein got his Nobel Prize for this rather than for Relativity Theory.

Bohr (1920s)- Single electron orbit description. Electrons are allowed certain orbits (aka **energy levels**). Remember, the classical question: **_Why don't electrons spiral into the nucleus if opposites attract?_** Good news: The orbit levels are integral, extending outward radially from the nucleus 1 through 7.

Beyond that, the electron is effectively ejected from the atom (think ionization).

Here we interpose the Rydberg equation: **ΔE** (absorbed or emitted per photon) =

$R_H (n_i^2 - n_f^2)$. For the above equation, **R = 2.18 x 10^{-8} J**. That's useful to calculate the energy differences between these "allowed" energy levels.

*You may have seen a different form of the Rydberg equation, i.e., beginning with "1/λ," so the **R value may vary**. However, regardless of which form you use, you should obtain the same numerical results for wavelength and energy.*

Here is a kind of thought experiment. For the moment, engage BOTH sides of the brain on this one. Imagine that we're dealing with only one electron for the moment. It's going to rise and fall between those levels we talked about.

OK, now calculate the energy and wavelength of each transition your instructor tells you.

NOW, come back closer to reality. Imagine that we have a gas sample of trillions and trillions of molecules or atoms. Their energies are rising and falling constantly, so you can and should imagine how many types of transitions that occur at any given moment!

This btw gives rise to the line spectra associated with the Bohr view of the atom. Each substance will give a unique line spectral pattern.

Mathematical summary for this part of the chapter:

$$\Delta E = R_H (1/n_f^2 - 1/n_i^2) = h\nu = hc/\lambda.$$

BE ABLE TO CALCULATE THE WAVELENGTH IN NANOMETERS AND THE ENERGY IN kJ/MOL OF ANY TRANSITION!

DeBroglie (1920s): Just as we now accept that light has both wave and particle properties, De Broglie asserted that ***electrons***

62

have not only particle but wave properties also. The problem up to that point was, we knew **how** electrons interacted with energy (digital/quantum) to maintain orbits, but we still didn't know why. The "why" of it all even eluded the astonished Bohr.

But let me give you the essence of this. The kind of wave we're talking about is a **standing/self-sustaining, self-reinforcing wave.** Get hold of a guitar or go online for a video. Watch a string being plucked and observe the vibrational pattern. The wave **"sustains"** for a while, although of course it eventually decays. In any event, you get the idea. Google "standing waves," and read about **constructive interference** of waves in your textbook.

I know, I know. It actually still doesn't fully explain sustained electron orbits, but trust me, it's the best explanation humanly speaking.

Your text gives you the matter-wave equation:

$$\lambda = h \,/\, (mass \; x \; velocity)$$

That's about as far as I go with my students. I just encourage them to plug in some macro numbers (i.e., pretend we're dealing with a baseball traveling at about 90 mph) and some subatomic ones (i.e., an electron traveling at around 2.57×10^7 m/s). They (and you) would see that in the macro situation the wavelength would be insignificantly small due to the huge mass, but that in the subatomic situation it would become significant.

Heisenberg (1920s) – The **Uncertainty Principle**: We can't know for certain both the position and the momentum of in this case the electron at any given time. The equation is:

$$\Delta x \,.\, m\Delta v \geq h/4\pi$$

DO NOT be intimidated by that equation; it just says that there will be an uncertainty of at least **h/4π** when we're trying to completely find out the whereabouts of the electron, and that value as you can see is a **constant**. This actually should sit OK with you, because life itself is never a certainty (*aren't I the* *great philosopher!)*, and we can assign a probability to pretty much every life event.

Schroedinger (1920s): Wave and probability functions and equations. This can get rather technical, but suffice to say that from these we distill four **quantum numbers**, symbols that indicate the unique address of each electron in the atom. Physically, these translate to three-dimensional spaces, called **orbitals**, where we are likely to find the electrons.

If you look at an orbital energy diagram of any atom except that of hydrogen, you will see that it gets rather complex. Basically, the reason for that is when you start adding even ONE electron, i.e., helium, where we have two electrons, the electron interactions get rapidly complex, and energy levels undergo splitting. The result is multiple energy levels and sublevels, as your textbook no doubt explains in detail.

Anyway, the first three numbers pertain to the orbital, or space, as I mentioned above. These numbers are:

Principal (n): Integers, ranging from 1 to 7. 1 is closest to the nucleus, and the rest expand radially outward. Main levels of electrons.

Angular, (l): Integers, ranging from 0 to n-1; n-1 is the **maximum**, not necessarily the only, value of l. Sublevels of varying energy. Some explanation: Within each principal level, you have "n" Angular sublevels. For example, in Principal level (shell) 1, we have one Angular level, and its symbol is $1 - 1$, or 0 $(n - 1)$.

The 0 is a symbol for the **"s" orbital**, which is **spherical** around the nucleus of the atom.

Magnetic (m): Integers, from $-l$ through $+l$, passing of course through 0. The magnetic quantum number is a symbol denoting **physical orientation**. Use a number line to help you keep track of these.

Example: If n = 1, l only has one value, namely 0, so the only magnetic quantum number at this level is 0 also. That makes

64

sense, as there really is no specific orientation for the s orbital other than spherically around the nucleus.

But at the Principal level n = 2, we have two l sublevels, 0 and 1. You already know that l = 0 signifies an s orbital. What about l = 1? Think **p orbital**. Actually, the p orbitals come in a set of three. To prove this, what are the m, or orientation, values when l = 1? Why, -1, 0, and +1. How many values would that be? Three! P orbitals are kind of dumbbell shaped, each 90° to the other two; think x, y, and z axes from your math class.

I'll let you do n = 3 on your own. You should come up with five orbitals. The shape of each of these is fairly complex (f orbitals are even more complex!), so suffice to represent them by five dashed lines at the same level. I suggest you Google them to see the shapes.

$$\mathbf{d} \; \_\_ \; \_\_ \; \_\_ \; \_\_ \; \_\_ \quad \mathbf{or} \; \_\_ \; \_\_ \; \_\_ \; \_\_ \; \_\_$$
$$\mathbf{d_1 \; d_2 \; d_3 \; d_4 \; d_5}$$

To reiterate, an **orbital** is a region where we are **likely** to find an electron, not where we DO find it. Remember from your textbook: we have to deal with **PROBABILITIES** from here on in. Each orbital (dashed line, or, if you like, a box) can hold a maximum of two electrons.

The fourth quantum number pertains to the electron itself.

Spin: The spin orientation, clockwise or counterclockwise, refers to the electromagnetic properties of the electron. Sometimes you will encounter the terms "spin up" and "spin down." The only way we can pair up two electrons within an orbital is if the spins are complementary, or opposite. Otherwise the repulsive energies would dominate and destabilize the arrangement.

I liken the first three quantum numbers to the construction of a multistory hotel or apartment complex. We establish the floors and rooms, and now we're going to place people on and in them. You just need to know the principles/rules for doing that.

Aufbau Principle: **The most stable configuration results under the current conditions.** We're going to place electrons in orbitals in order from lowest to highest energies. That should make sense. So, what exactly is that order? We'll see shortly. The "observed" order of filling has a handful of exceptions to what we would *intuitively* think, **but these exceptions reflect the greater overall stability of the configuration in the end.**

a) **Pauli Exclusion Principle**: Each electron has a unique address. Oh, I know, your textbook says each electron has a unique set of 4 quantum numbers. Same thing, really. Within each orbital, remember, the electrons are distinguished by their (opposite) spins.

b) **Hund's Rule**: In a set of orbitals of the same energy, such as p and d, electrons go in singly with the same spin before they pair up. Look at it this way. Round table, several card players, including the dealer. The dealer gives each player one card first, face down, then a second card in the same fashion but face up. You get the idea. Why? It's what we observe, so the system must be more stable that way (it is!).

If there's at least **one unpaired electron** when we're done, the element is **paramagnetic**. If all electrons are paired, it's **diamagnetic. Look up these terms and be able to describe the behavior.**

Now, I can help you write out ground state electron configurations, but you will find out yourself that it's very easy. Like anything else, it just takes practice, so write out a few daily.

Make sure you have a Periodic Table with you when you write these. You should be provided one during a quiz or test, by the way.

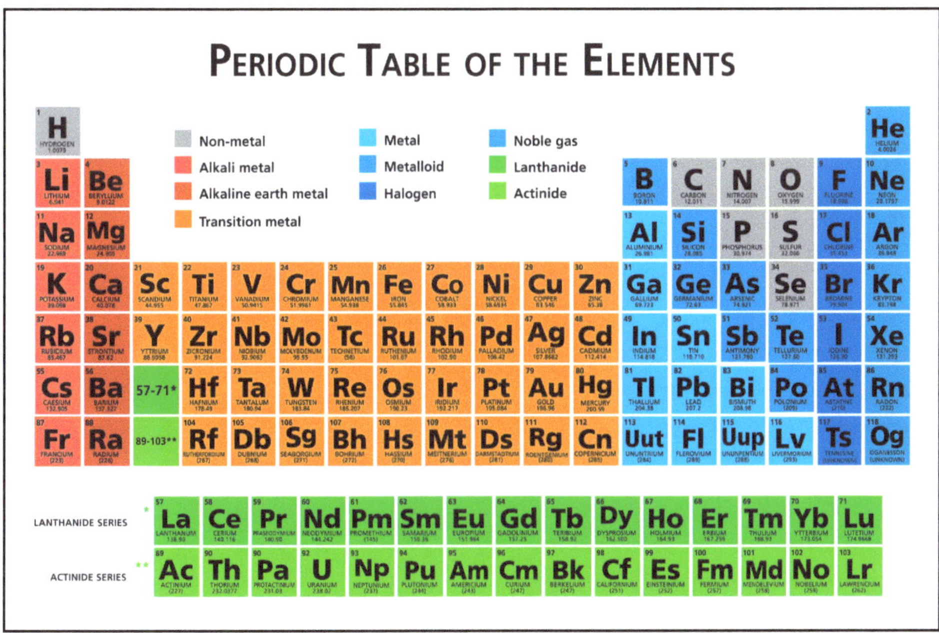

The first two groups (shaded in red) are called the "s" block (why?), the groups shaded in blue the "p" block (again, why?), the groups shaded in orange the "d" block, and lastly, the parts of Periods 6 and 7 shaded in green the "f" block.

Let's start with hydrogen. 1 (Principal level), s (lowest energy sublevel), and finally

^{1}(superscript for the number of electrons in the orbital), $1s^1$.

With helium, the 2nd element, just add an electron to fill the 1s orbital, $1s^2$.

Now, if you look at the Periodic Table and advance per the Aufbau Principle, with lithium you need to start a new Principal level, 2, so the configuration would be $1s^2 2s^1$.

A little further down, e.g., at carbon, the configuration would be

$1s^2 2s^2 2p^2$. Just so we keep the right visual images, what would the energy diagram for that configuration look like?

After a while, with large numbers of electrons, it can get tiresome writing out the configurations longhand. Learn the shorthand (**look it up**) and be SURE to write the format correctly. Over the years many test points were lost because students did not keep track of formatting.

Post-script of this section: You probably know by now that the **4s level is slightly lower in energy than that of the 3d.** For example, the last electrons in K and Ca are $4s^1$and $4s^2$, respectively. There are NO 3d electrons until scandium, element 21. Again, why is this the case? The simplest answer is that with multielectron atoms, the electron-electron interactions become impossibly, inconceivably complex, as mentioned earlier, and energy levels get "split" in virtually unpredictable ways. We accept it because that's the way it is. But use your Periodic Table to help you determine the order of filling, and I can virtually guarantee you will never go wrong writing out these configurations.

You probably by now know of some exceptions to the order of filling. We have a handful of examples in the Periodic Table. For example, the last electrons of Cr have the configuration $4s^13d^5$, rather than $4s^23d^4$. Why is this so? Your immediate response should be "because it must be more stable!" Of course! The **PRINCIPLE** of lowest to highest energy order has not been violated. It turns out that due to the closeness in the energy of the 4s and 3d levels AND the number of electrons involved, there MUST be a greater energy cost to pair up the 4s electrons and leave one 3d orbital vacant, versus **single electrons with parallel spins in each of the orbitals** (see your text or Google).

Writing ionic electron configurations

There are just one or two points to be made here. Obviously, a cation will have fewer electrons in the configuration than the parent atom, and the anion will have more. What I advise my students to do is to write out the configuration for the neutral atom first. Then remove or add electrons to get the final configuration. In the case of s or p block ions, most of the time you will arrive at the nearest noble gas configuration because you are either removing s electrons (metals) or adding p electrons (non-metals). By the way, you should take note of that trend: **An atom loses or gains electrons to**

form the familiar ion which coincidentally has the electron configuration of its nearest noble gas neighbor. Hmmm...... driving force?

Shielding

Simple rule: **Inner electrons shield outer electrons from the full pull of the nuclear charge.** Think about it......electrons repel each other, and since the inner electrons were there first, the outer electrons kind of lose out on the full protonic attractive force. Since the force is radial and the inner electrons form a narrow circle around the nucleus, the outer electrons will be held less tightly. So, the actual pull felt by any given electron is termed the **effective nuclear charge** or the **net nuclear charge.**

This rule, however, does not let you off the hook; READ the full discussion in your text about p electron penetration of the electron cloud, etc., and pay attention to your instructor's discussion as well!

The Periodic Trends

General principle: There are several periodic trends you must know, and the source of these trends pretty much boils down to a battle between nuclear charge and shielding effects.

Just a bit of history: There are two key figures you'd better know about- **Mendeleev** and **Moseley**. They're easy to remember. Both names start with M. Mendeleev made his mark in the 1800s, and Moseley in the early 1900s. Also, they're easy to distinguish: Mendeleev organized his Periodic Table in terms of increasing **atomic mass**, and Moseley organized his in terms of increasing **atomic number**. Big difference, because the periodic repetition of properties is smoother with the latter. To Mendeleev's credit, his was really the first rigorous organization of the Table we use today, and it actually predicted the existence of elements theretofore unknown.

Moseley used x-ray spectral analysis to resolve the apparent discrepancies of the handful of elements that didn't fit with Mendeleev's organizational pattern. His **modern periodic**

law, still in use today: **The properties of the elements are a periodic function of their atomic numbers**.

The **trends** of the following **properties** roughly follow a northeast/southwest, or diagonal, path.

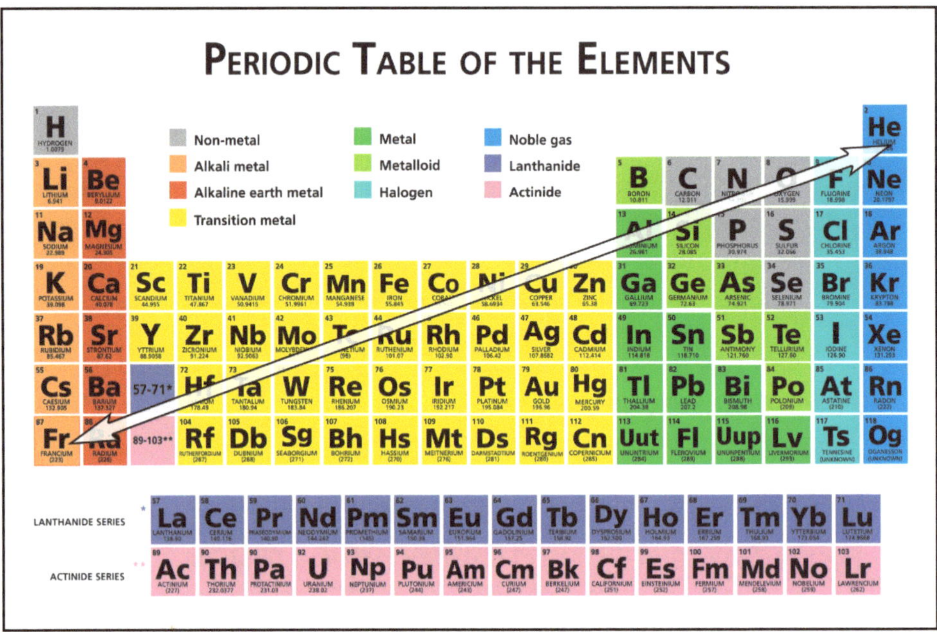

Atomic and Ionic Radii:

As we go northeast, radii decrease. So, what happens as we go southwest? Let's break this down into two components.

For atoms, as we go across the Table, the nuclear charge is increasing **WITHIN THE SAME PRINCIPAL QUANTUM LEVEL!** So even though electrons are also being added, the shielding effects are overall less than the effect of increasing the nuclear charge. As a result, the atomic size shrinks. As well, as we go north, the principal quantum shell reduces by one each time, so the shielding is drastically reduced. Of course, the number of protons drastically reduces, but the overall volume shrinks, so the atomic size does naturally.

But, you say, **what about ionic sizes**? In context, the same trend occurs. For positive ions, as we go across a period, the ionic charge (+) goes up, for example, from Na+ to Mg^{2+} to Al^{3+}, for example, and the number of **protons** simultaneously goes

70

up. A double whammy! The ionic size shrinks! That should make immediate and thorough sense to you! The same occurs in the anion camp. For example, N^{3-} means that poor nitrogen is saddled with three extra electrons, and is rather swollen compared to its neighbors O^{2-} and F^-, where the number of protons increases across the period. Note the simultaneous reduction in the negative charge by a unit. See the trend? Also, as with atomic size, as we go north, the number of principal shells reduces.

If you understand those trends, the others should be easy to understand as they follow from the charge/shielding arguments.

Ionization Potential (Energy)

It takes energy to remove the outer electron(s) from the atom, and the more tightly they're held (smaller radius, remember?), the more energy is required. So "I.P." goes up as we go northeast, and vice versa. Make sure you can explain this in your own words.

Electron Affinity

In plain English, how amenable is an element to acquiring additional electrons? This trend is a less smooth transition, but over all it follows the others. It rises as we go northeast, and vice versa.

Metallic Behavior

This time, such behavior **increases** in a southwest direction, where electrons tend to be lost relatively easily.

When I teach the above trends, I tend to de-stress compartmentalization in favor of the continuum approach. For example, from northeast to southwest, all elements will exhibit metallic and nonmetallic "behavior" of continuously varying degrees.

Warning: You may be asked to compare the magnitudes of any of the above properties between two elements.

Post-script: By now, you should have more finely organized your Periodic Table in terms of not only physical but now chemical properties, such as denoted by **s block**, **p block**, **d block**, and **f block** elements.

In summary, be able to: a) write configurations "on sight," b) describe and explain trends, and c) compare two elements in that context.

The Periodic Table is the cornerstone of chemical property and behavior! Learn to lean on that Table (no pun here). You will be supplied a Table for each quiz or test, and the more you understand its organization, the less you'll need to memorize!

Bonding

Remember the Dalton Dance? All the stuff you've learned so far converges onto the topic of **chemical bonding**. After all, mainly what one does in chemistry is observe chemical reactions. Guess what they involve? That's right, **breaking and making chemical bonds**.

There are likely two chapters in your textbook about bonding. According to your textbook, there are two categories: **ionic** and **covalent**. However, **I mainly teach bonding as a continuum; ionic bonding is an extreme case of covalent bonding, or electron sharing.** That sounds heretical, because your lecture instructor is going to tell you that ionic bonding means rather than sharing, electrons are actually being transferred completely from one atom to another.

But if you think about it, the two atoms will be associated anyway in the ultimate formation of the ionic bond. What I think is amusing is that almost immediately after your book gets done delineating the two types of bonding, it discusses ionic and covalent **"character"** of bonds. So I will boldly restate it: **Ionic bonding is merely an extreme case of unequal**

valence electron sharing. By the way, if you're planning on taking organic chemistry after this, it's rather likely your instructor will make frequent references to the "character" of a bond.

Now, remember, why do substances react in the first place? Things are more stable on the other side of the arrow.

This is particularly important when your instructor covers the topic of lattice energy. Ionic compounds form rather ordered crystals made up of alternating cations and anions. The stability of these crystals, called **lattice energy**, is directly dependent on the product of the ionic charges and inversely proportional to the square of the distance between their nuclei. Therefore, **small, highly charged** ionic arrangements are more stable than their larger, less charged brethren. By the way, **charge has more influence than size variations**, so you probably don't need to split hairs in comparing the lattice energies of ionic compounds. Make sure you are able to make reasonable comparisons of compounds vis-à-vis lattice energy (stability).

A word about the **Born-Haber Cycle**: Your instructor might ask you to learn this, and if so, my sympathies. To pay brief homage, it is useful to know as the source for determining lattice energy.

Now that you know about electron configuration, the impetus for things to react is that atoms want to acquire a valence (outer) electron configuration that "looks like" that of their **nearest noble gas neighbor**. So. an atom will lose, gain, or share valence electrons to achieve that. As you know, metals tend to lose electrons and non-metals tend to gain them.

Polarity and Electronegativity – critical bonding concepts

Electronegativity is the attraction that an atom has for electrons **WITHIN A BOND**. Those last three words are extremely important. Now that we're talking about bonding, we assume that the atoms are now not alone.

You have an **electronegativity chart** in your textbook. Stop reading and look at it for a minute or two. OK, do you see a trend? Right; as we go **northeast**, the electronegativity **rises**, and vice versa. Very similar to the earlier trends we saw! Actually, it's a remarkably smooth trend, isn't it? Quickly right now, pick any two elements. For the moment, we'll pretend they're bonded. Calculate (in your head, for heaven's sake!) the absolute value difference between the two. If you get a number other than zero, technically you have a **polar bond**. The higher the number, the more polar the bond.

Polarity is a central concept in chemistry, and is of interest particularly where reactivity is concerned. In a polar bond, the bonding electrons on average lie closer to one atom versus the other. That gives that atom a slight negative charge and the other atom a corresponding positive one. Don't worry about the magnitude, by the way. As long as we have a bonding situation of unequal electron sharing, we have by definition a polar bond.

Now, your textbook says that if the electronegativity difference is equal to or greater than some number, e.g., 2.8, the bond is essentially ionic. Well, well. **Do you see why I talked about bonding as a concept of degree and not of kind?** Look at the polarity "chart" below.

BOND POLARITY

0%--100%

0% might be thought of as reflecting a pure covalent bond (equal electron sh4aring), while at 100%, yes, electrons are transferred and the bond is ionic. But that means that in between, we have a rather large spectrum of **bond "character,"
ionic and covalent**. For example, if the bond polarity is 20%, what does that mean? One way of interpreting this is **80% covalent character**, and **20% ionic character.** Yes, the electrons are shared, but unequally. This is rather important, as it will help us reasonably predict, for example, some physical properties as well as the reactivity of the compound in question.

74

UNDERSTANDING BOND POLARITY IS CRUCIAL TO SUCCEEDING IN ORGANIC CHEMISTRY!

The **Noble Gas Rule** (a step up from the **octet rule**): In a chemical reaction, an atom will lose, gain, or share electrons so as to have a **valence electron configuration of its nearest noble gas neighbor**. Read this carefully; it does NOT mean, for example, that when sodium is ionized it becomes neon. Rather, its **valence shell becomes identical to that of neon**, which of course represents relative **stability**.

The same rule applies to more covalently inclined reactions. Just remember that when electrons are shared, all become "community property." Once again: Why are the atoms becoming molecules in the first place? The end result is more stable!

Here's my summary of how to write Lewis structures in general.

1. Add up all the valence electrons of all atoms involved. If you're dealing with a negative polyatomic ion, add in the extra electrons according to the charge on the ion. If the ion is positive, subtract that number from the total.

2. Write a reasonable skeletal structure for the molecule or ion. Generally, the less/least electronegative element goes in the center. Some textbooks use the phrase "the element with the fewest number of valence electrons." Don't freak out; sometimes the only other element you're dealing with is hydrogen (i.e., ammonia, water, etc.). Obviously, since hydrogen only bonds once, it can't go in the center!

3. Create single (2-electron) bonds, and as necessary strew the remaining electrons (non-bonding) around the atoms to give each a total of eight. Many times this will use up all the electrons you totaled up in step 1. In that case, "If it ain't broken, don't fix it." Up to this point, you've reached the goal: a plausible Lewis structure.

4. Adjust as necessary. That's a loaded statement. Often somebody lacks eight electrons, so you need to move electrons around the "hacienda" and create double (or

triple) bonds so everyone does in effect have eight. I think you see what I mean. Like anything else, it takes a bit of practice.

**PLEASE, PLEASE, I BEG OF YOU!
MAKE SURE THAT BEFORE YOU LEAVE OR
SUBMIT ANY LEWIS STRUCTURE YOU VERIFY
THAT YOU HAVE NO MORE OR LESS ELECTRONS
THAN YOU STARTED WITH, AND THAT ALL ATOMS
HAVE EIGHT ELECTRONS AROUND THEM!**

Having said that, if you haven't gotten there yet, your textbook says that the noble gas ("octet") rule is the first thing to go out the window. Look at all the exceptions!

That's just great, Dr. C. Now what do I do?!

RELAX! Take a deep breath, and lean on the **fundamental stability principle**. Things react to form the **most stable structure**, configuration, etc.

We have to look at the data and analyze why. For example, why do PCl_5, $BeCl_2$, SF_6, to name a few compounds, have central atoms with either **less** or **more** than a noble gas valence electron configuration? Your answer should be, without hesitation, **THEY MUST BE MORE STABLE THAT WAY**! Remember, I said way back at the beginning that this is a perfectly reasonable and actually a very wise response! OK, ok; a more satisfying answer from a mechanical standpoint is that **below Period 2, d orbitals and their electrons become available for bonding.**

Some words about **resonance** and **formal charge**: These **bookkeeping methods** are introduced to help you write the **most plausible Lewis structures**. Think of them as refining tools.

I don't particularly like the textbook definitions of resonance, because often they don't actually define.

Resonance is a **phenomenon characterized by a delocalization of some electrons in a molecule or ion.** The purpose of

76

resonance, or **electron delocalization**, is, you guessed it, to make the molecule or ion **more stable**. Remember, electrons tend to repel each other, and if there is some legitimate way to spread them out, that is favored over confining them.

Typically, when checking a structure for resonance, the presence of alternating double and single bonds suggests resonance is present. Now, that doesn't mean that if you see a double bond it automatically means that resonance is present. Make sure there are plausibly alternate ways of writing the structure.

The problem is that the Lewis structures are not accurate in illustrating reality. Here's where left-brain/right-brain thinking comes into play. I have my students imagine taking the various resonance structures of a molecule/ion and placing them all into a "blender," and then imagining what a resultant weighted average "hybrid" would look like. There's where the artistry applies. **The limitation in illustrating the structure is ours, not the molecules**.

Next time you make a fruit smoothie, just before you blend all the ingredients, inspect each one for quantity and quality (i.e., color, texture, etc.). Then before you blend all the ingredients, **imagine** what the weighted average would look like. Get the idea? It's a good juncture to remind you that you'll never fully understand chemistry unless you **regularly engage the artistic side of your brain.**

Examples where resonance is operating: O_3, CO_3^{2-}, NO_3^-, SO_3, PO_4^{3-}, and SO_4^{2-}. As your text says, draw the various resonance structures. In each case, all these structures are **collectively** an approximation of the actual structure. With the help of **formal charge** (see below) rules and calculations, I'm confident you'll hone in on the optimum single Lewis structure. Then, as a final touch, I encourage you to use your artistry to experiment with drawings to draw the "weighted average" Lewis structure.

Post-script: As I hinted above, resonance will be tremendously important in understanding **organic chemistry**. Take good notes and file this information away if you are planning on taking that later.

BTW, if your instructor happens to catch you drawing that average structure, just explain that you're getting ready for organic chemistry!

Regarding **formal charge**, as stated above, this is merely another **bookkeeping device** to give you a rough (and I mean **rough**) idea of where the valence electrons are. Of course, its use is to help us, along with resonance, write the most plausible Lewis structure. However, this ironically emphasizes the **limitations of Lewis structures**, as you saw earlier.

With formal charge, I don't want to complicate the issue here.

Sorry, Dr. C., that ship has sailed.

Three things to keep in mind when you analyze.

1. Remember, the actual structure will be a hybrid (weighted average) of all resonance structures.

2. Numerically lower charges are better than higher ones.

3. More favorable structures will place any negative charges on more electronegative atoms.

By the way, you should notice by this time that the "noble gas rule" has not only gone out the window, it's probably hundreds of miles away.

Bond Energy

I sound like a broken record, but here we go again: In a chemical reaction, bonds must be broken and reformed in a different way so that new substances can be formed (remember the Dalton Dance?). Since you've already been through the agony of the chapter on thermochemistry, why put you through it again?

Ergo, I merely tell my students that the old familiar **ΔH of reaction** they've already seen (*heat of the products minus heat of the reactants*) turns up here as the mathematical difference between the **energy of the bonds broken and the energy of the bonds formed**. It remains to be seen, of course, whether the sign of that whole mess will be negative or positive.

78

I do introduce the concept of **average bond energy per mole** (see the chart in your book). But I've decided it's not worth going through the agony of the actual computation.

However, be ready to go through this if your instructor requires it! If that's the case, again, you have my sympathies!

Molecular Geometry (Shape)

Molecules are 3-dimensional (surprise!), so it makes sense to talk about them that way. With molecular shapes, we really just take a 2-D Lewis structure and transform it into a 3D picture by using some basic geometry knowledge you probably learned in grade school. Then we add some basic electrostatic knowledge. Vo-la! A complete molecular model. By the way, all the better if you can get hold of a **molecular model kit** or make models from Styrofoam balls and toothpicks, for example.

We'll use the **VSEPR theory** your instructor is no doubt discussing.

1. Stick with the **valence electrons** we used for Lewis structures. So, start with the most plausible Lewis structure when going from 2D to 3D.

2. We'll focus on the **central atom** you used when you wrote your Lewis structures. The central atom will be the reference point. In particular, what types of electrons do we have around the central atom?

1. We'll talk about **regions of electron density** around the central atom. There are two types of electrons here: **Bonding electron pairs** (BP) and **lone electron pairs** (LP). Each lone pair of electrons is considered **one region**. Bonding pairs of electrons are considered one region as follows: A **single bond, double bond, and triple bond are each ONE region. They are all equivalent**. Why would that be? Bond pairs of electrons are constrained between the two nuclei of the atoms of the bond, so we don't distinguish among single, double, or triple bonds.

2. This is **VSEPR**, so remember that electron groups tend to **repel** one another, and according to our basic principle, the **most stable geometry will prevail**. There is a hierarchy of repulsive forces:

- **LP – LP > LP – BP > BP –BP**.

You can memorize this, but it's better to understand how the above comes about. Lone pairs love to spread out, so they form huge bubbles of electronic volume. That's why two lone pairs of electrons, as in water, for example, tremendously repel one another and as a result suppress the H-O-H bond angle to well below the idealized 109.5°. Keep that in mind. The bonding pairs don't stand a chance because they're constrained already. Do you see the difference? It's a dog eat dog world out there in electron land!

3. There are three basic geometries we use:

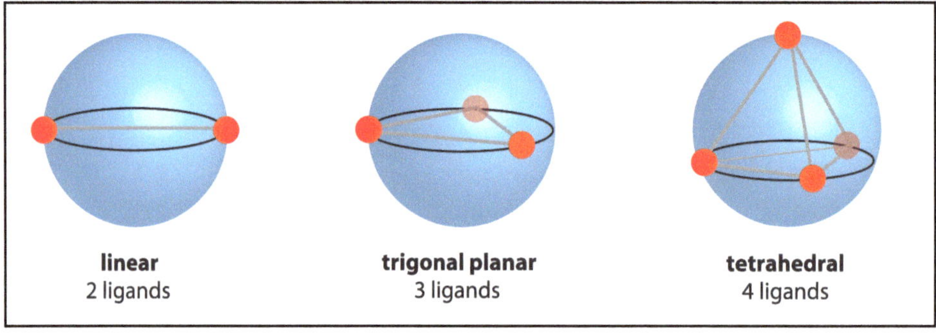

All the other molecular shapes you encounter are pretty much **derivatives** of these three. Notice below that I included some basic bond angles. These should come as no surprise. For example, two distinct points determine a line. What's the angle? Why, 180° of course. And so on.........

Regions of electron density around the central atom	Basic Geometry (electronic)	Central bond angle (degrees)
2	Linear	180
3	(Trigonal) Planar	120
4	Tetrahedral	109.5
5	Trigonal Bipyramidal	120 (horizontal) 90 (vertical)
6	Octahedral	90 (all)

The *Molecular (Final)* versus the *Electronic* Geometry

When assigning a final shape to your molecule or ion, by convention we "erase" the lone pairs around the central atom. What's left?

Obviously, if we're dealing with all bonding pairs, nothing changes. Lone pairs are another matter. For example, water has an **electronic** shape of a tetrahedron, but it ends up being called *angular* or *bent* when we **factor out the lone pairs**. Make sense? By the way, ammonia ends up being called a *trigonal pyramid* when we "erase" the lone pair.

With practice, you'll become more familiar with the derivative geometries and be able to draw more examples. **Remember, I'm not trying to re-invent the wheel here**. Your textbook has wonderful (no doubt computer-generated) illustrations in ample quantities.

Two Post-scripts

1. Lone pairs of electrons **WILL suppress the central bond angle** below the **idealized** angle, and your instructor may want you to memorize these modified angles.

2. VSEPR does a good job of explaining what we see on the spectroscope in terms of shape, but we still have not explained **bond lengths** or **bond energies.** For example, we know that the shape of methane, CH_4, is a perfect

tetrahedron, with all bond lengths and angles the same. VSEPR allows us this perfect shape, but really doesn't explain it. We need to invoke another **theory** to **explain** what we see spectroscopically.

Molecular Polarity

You should now have a solid knowledge of the basic concept of polarity. Regarding **molecular** polarity, however, you will need to keep two factors in mind in your analysis: **electronegativity** and **geometry** (shape). As well, I want you to invoke your overall intuitive sense of balance and symmetry, and you then should be able to INSTANTLY determine whether a molecule is polar.

I know, I know, your textbook discusses force vectors and dipole moments, but for our purposes at least, you won't need to invoke these in your analysis.

Draw the following molecules taking into account geometry as well as Lewis structure (please!): **methane** and **dichloromethane** (CH_2Cl_2). Oh, come on, of course you can!

Now, what is one obvious difference between them?

Of course! The dichloromethane has two chlorine atoms substituting for the hydrogen. Which one do you think is polar? It should be obvious that the dichloromethane is. Why? Methane is nice and symmetrical, but in dichloromethane the chlorines throw everything off! We might say that the electronegativity is greater with chlorine than with hydrogen, **but we also need to include the geometric factor**. With only the Lewis structure, the compound might look planar:

82

$$Cl \!-\! C \!-\! Cl$$

with H above and H below the carbon.

The Lewis structure alone makes the compound look symmetrical, but in a tetrahedral pattern, according to VSEPR rules, the **actual geometry is tetrahedral**. I think you get the idea. Now pay attention to the other examples in your textbook.

(**draw a tetrahedron with each of the atoms at the apices and the carbon in the center.)

Valence Bond Theory

When two hydrogen atoms get together to form H_2, the initial attraction is between the electrons and the opposite protons (nuclei). They can only approach to a certain optimum distance; if the atoms get too close, the protons begin to repel one another, as do the electrons.

In the above example, we **explain** by saying that **the two 1s orbitals overlap in an optimum fashion to give an energetically favorable molecular bond.** In another example, when HCl forms, **a 3p orbital from chlorine overlaps with a 1s orbital from hydrogen** in optimum fashion. The bonding in HCl is better explained by using the orbital overlap than simply by Lewis structures because of the different **orbital sizes** and the resulting **bond length and energy**. Of course, the valence electrons pair up for a single bond, as before, hence the **Valence Bond Theory**.

REMEMBER: WE EXPLAIN THE EVIDENCE IN NATURE. WE DON'T CREATE IT!

Now, in many molecules, even simple orbital overlap cannot explain observed bond length and bond energy. In our methane example, **spectroscopic evidence tells us that all CH bond lengths and bond energies are the same.** However, the valence electron arrangement would indicate that both **s and**

p orbitals of carbon are involved in the bonding. That would mean in some cases, the s orbital of carbon would overlap with the s orbital of hydrogen, and in others the p orbital of carbon would overlap with an s orbital of hydrogen. Try to picture the resultant shape! Utterly lopsided and awkward! Most importantly, however, it is **NOT what we actually see on the spectroscope! What we see are four equal bond lengths and equal bond energies!**

Hybridization

What we think happens (theory/mechanism) in the methane case is that just as in biology, where DNA is mixed, orbitals such as s and p are mixed in a sort of quantum mechanical blender to yield a set of *hybrid orbitals*, **all with the same energy, and of an average energy intermediate between that of a pure s and a pure p orbital**.

Stop right now and imagine that scenario, as before.

Think of it as mixing coffee and milk. The entire mixture will be an intermediate color between that of pure coffee and pure milk. We can vary the shade by changing the relative proportions of coffee and milk. You get the idea.

We can mix, for example, the 2s orbital of carbon with one, two, or all three of carbon's 2p orbitals and come up with a set of orbitals with a shape and energy intermediate between those of the s and the p alone. Additionally, for carbon, we promote one of the 2s electrons to the vacant p orbital.

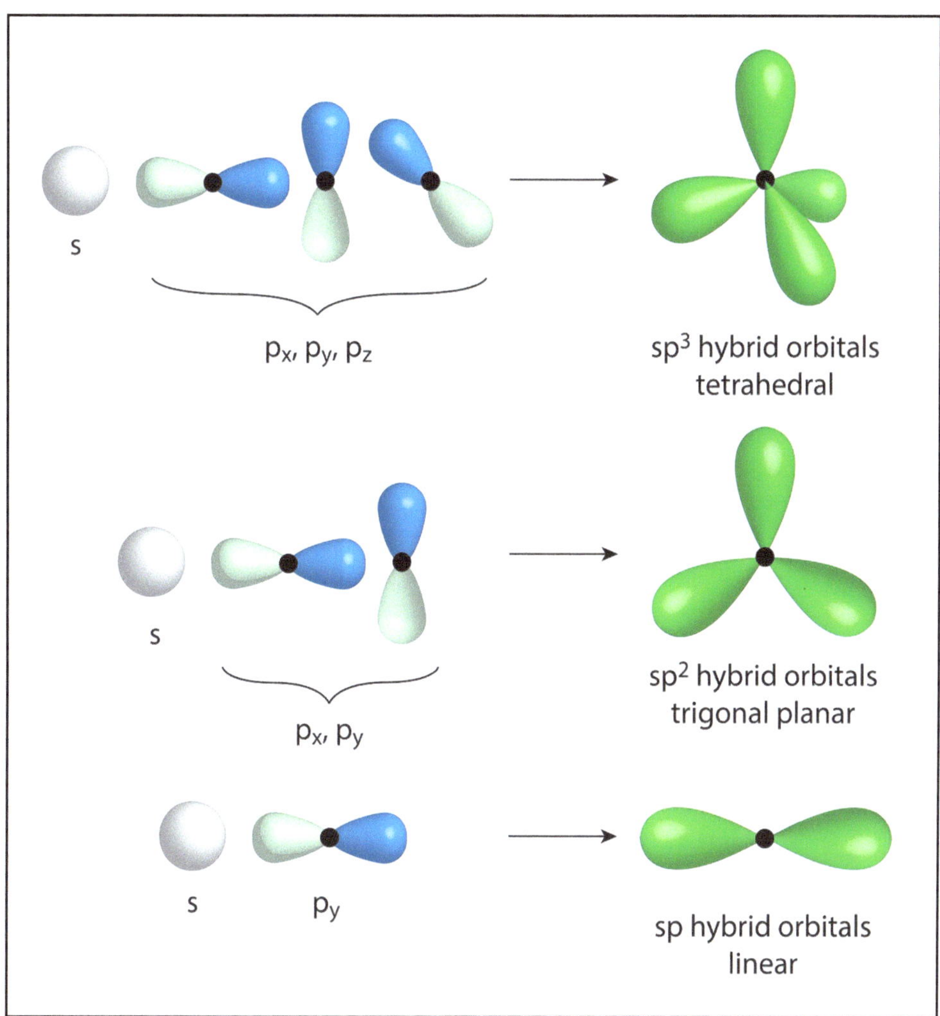

Conservation laws indicate that we end up with the same total number of orbitals we started with.

The only things left to explain are the nuts and bolts of the theory!

For example, as I stated earlier, for methane, we **observe** (law) that all four CH bond lengths and energies are equal. How do we **explain (**theory) this?

For methane, we mix **an s** and **three p** orbitals in our **quantum mechanical blender** to produce **four new** orbitals of **weighted average/hybrid** length and energy, each with an appearance somewhere between that of the **s** and of the **p** orbitals. As the artist in you might speculate, the hybrid shape would be closer to that of a pure p rather than of an s orbital, since we are mixing three **p's** with only one **s.**

The orbitals point to the four corners of a tetrahedron to minimize the repulsions among the electrons that occupy them, per the diagrams above. They each overlap with a 1s orbital from hydrogen to form the four equivalent bonds we see in methane. We call these "sigma" bonds; they're essentially single bonds characterized by a straight-on, head to head overlap of the orbitals between the nuclei of the two atoms. In methane, we have **four sp^3-s sigma bonds**. Voi-la!

Now, it's a short step to show how **sp^2 hybridization** explains the planar geometry and how **sp hybridization** explains linear geometry. **Sp^2 hybridization** is the result of blending **one s** and **two p** orbitals and leaving **one p orbital pure**, or unhybridized. So, we have three new orbitals that point to the three corners of a plane, with the pure p pointing 90° above and below that plane. With **sp hybridization**, that's right, only two orbitals, **one s and one p**, are mixed, leaving **two pure p orbitals** that are 90° to each other and to the line formed by sp hybridization.

Regions of electron density	Basic Geometry (electronic)	Basic bond angle (degrees)	Hybridization of the central atom
2	Linear	180	sp
3	(Trigonal) Planar	120	sp^2
4	Tetrahedral	109.5	sp^3
5	Trigonal Bipyramidal	120 (horizontal) 90 (vertical)	sp^3d
6	Octahedral	90 (all)	sp^3d^2

As I mentioned earlier and as your text probably says, **d orbitals can be used with certain elements below Period 2.** By that time, the orbital energies are close enough that the d's can be recruited.

Note: Don't throw out VSEPR; we still need electron repulsion to help explain the resultant shapes and central bond angles.

Whenever you hear about **pi bonding**, think double and triple bonds. As your text explains, we typically use **pure** (unhybridized) **p** orbitals which are 90° to the sigma bonds, and they overlap side by side in two places, above and below that sigma line. The overlap is rather **weak,** which accounts for the reactivity of the double or triple bond in many cases. **You're going to get a flood of this in organic chemistry. Be ready!**

Points to Remember:

- Sigma bond:-single bond; between the nuclei of the two atoms involved

- - head to head overlap; relatively strong; makes up the molecular "skeleton"

- Pi bond:- two p orbitals bonded in a **side by side overlap**; 90° to sigma bonds; **relatively weak; indicates potentially reactive location**

Molecular Orbital (MO) Theory

MO theory picks up where valence bond theory leaves off, and it helps us explain certain cases of **paramagnetism** and **spectroscopic** data.

The real use I see in MO theory is that it goes a bit further than resonance in explaining electron delocalization and the resulting molecular stability. The "mechanics," if you will, involve combining **atomic orbitals (AOs)** from each atom to get **molecular orbitals (MOs), both bonding and antibonding**. The problem is, the MOs can't easily be visualized per se. One useful way of looking at MO theory is that it describes **orbital** (MO), rather than **electron**, delocalization, so that now orbitals become community property through atomic orbital combination.

Practically speaking, you're essentially stuck with some weird-looking diagrams and a set of rules regarding filling up the MOs with electrons. The real challenges are actually **drawing the diagrams and knowing about the orbital splitting**. For this you must look explicitly to your textbook and/or instructor.

However, once you have constructed the diagrams, simply use the **Aufbau principle** and **Hund's rule** to add the electrons in a vertical order. Then calculate the bond order, which is simply the difference between the number of bonding electrons and the antibonding ones, divided by two. As long as you obtain a number above zero, then in theory that molecule would exist. Sounds like a bookkeeping device, doesn't it? I say, let's look at the spectroscopic data and reason from there.

Appendix I – Polyatomic Ions

	-ide	hypo__ite	-ite	-ate	per__ate
Group 17-F	F^-	FO^-	FO_2^-	FO_3^-	FO_4^-
Cl	Cl^-	ClO^-	ClO_2^-	ClO_3^-	ClO_4^-
Br	Br^-	BrO^-	BrO_2^-	BrO_3^-	BrO_4^-
I	I^-	IO^-	IO_2^-	IO_3^-	IO_4^-
Group 16-O	O^{2-}	-	-	-	-
S	S^{2-}	SO_2^{2-}	SO_3^{2-}	SO_4^{2-}	SO_5^{2-}
Se	Se^{2-}	SeO_2^{2-}	SeO_3^{2-}	SeO_4^{2-}	SeO_5^{2-}
Te	Te^{2-}	TeO_2^{2-}	TeO_3^{2-}	TeO_4^{2-}	TeO_5^{2-}
[a]Group 15-N	N^{3-}	NO^-	NO_2^-	NO_3^-	NO_4^-
P	P^{3-}	PO_2^{3-}	PO_3^{3-}	PO_4^{3-}	PO_5^{3-}
As	As^{3-}	AsO_2^{3-}	AsO_3^{3-}	AsO_4^{3-}	AsO_5^{3-}
[b]Group 14-C	C^{4-}	CO^{2-}	CO_2^{2-}	CO_3^{2-}	CO_4^{2-}
[b]Si	-	SiO^{2-}	SiO_2^{2-}	SiO_3^{2-}	SiO_4^{2-}
[c]Group 6-Cr	-	CrO_2^{2-}	CrO_3^{2-}	CrO_4^{2-}	CrO_5^{2-}
[d]Group 7-Mn	-	MnO_2^{2-}	MnO_3^{2-}	MnO_4^{2-}	MnO_4^-

[a]Alas, of the nitrogen series in Group 15, only the monatomic ion has a 3- charge, while all the others have a charge of 1-. This is an exception to the pattern in Group 15.

[b]The first column in this row is blank because the monatomic ionic charge on Si is usually 4+

[c, d]The first column of these rows is left blank as the charges are variable.

Other polyatomic ions of interest:

C2H3O2- acetate, S2O32- thiosulfate

$C_2O_4^{2-}$ oxalate	SCN^- thiocyanate	$Cr_2O_7^{2-}$ dichromate
HCO_3^- hydrogen carbonate or bicarbonate	HPO_4^{2-} hydrogen phosphate	$H_2PO_4^-$ dihydrogen phosphate
HSO_4^- hydrogen sulfate	HSO_3^- hydrogen sulfite	
OH^- hydroxide	O_2^{2-} peroxide	CN^- cyanide
CO_3^{2-} carbonate	H_3O^+ hydronium	NH_4^+ ammonium

Appendix IA: Formula writing

Formula Writing of Ionic Compounds (Salts)

Ionic compounds consist of one or more positive ions combining with one or more negative ions to form an electrically neutral formula unit (ionic compound), referred to often as a salt. Ionic compounds typically contain at least one metal ion and a non-metal ion, i.e., $NaCl$.

Use the principle of electrical neutrality when writing formulas: The sum of all charges must equal 0. Write down the positive ion symbol(s) and charge first, and next to it the negative ion symbol(s) and charge. Then determine the smallest integer that is a multiple of each of the charge numbers, as in mathematics ("least common multiple"). Dividing that number by each charge number gives the number of each ion required. Write that number as a subscript after the ion symbol(s).

An equivalent and faster method involves simply "crossing over" the number of each charge to the other ion, writing that number as a subscript as above. For example, for iron(III) sulfate, we have Fe^{3+} and SO_4^{2-}. Crossing over, we have $Fe_2(SO_4)_3$.

Formula Writing (and Naming) of Acids

Acids, compounds with at least one H+ at the beginning of the formula, may contain polyatomic anions (i.e., H_2SO_4) or monotonic ones (i.e., HCl). In any event, they react with compounds called bases to form a salt and water. The base in most cases replaces the hydrogen ion (H+) of the acid with a metal cation, creating the salt, i.e.,

HCl + $NaOH$ → $NaCl$ (sodium chloride) + H_2O.

To find out the nature of the acid that made our salt, replace the metal ion(s) with the same number of H+. For example, for the compound Na_2SO_4, which is sodium sulfate, replace both Na+ with H+. You would then have H_2SO_4. "ATE" salts come from "IC" acids, so the parent acid is sulfuric acid. Similarly, "ITE" salts come from "OUS" acids. So if our salt is Na_2SO_3, which is sodium sulfite, we replace the Na+ with H+ ions, to get H_2SO_3, which is called sulfurous acid.

Other examples:

HNO_3 nitric

HNO_2 nitrous

H_2SO_3 sulfurous

$HClO_4$ perchloric

$HC_2H_3O_2$ acetic

If the anion of the acid is monatomic, insert the prefix "hydro" and end with "ic" invariably.

HCl hydrochloric acid

HBr hydrobromic acid

H_2S hydrosulfuric acid

Appendix II - Types of Equations

<u>**Combination/composition/synthesis**</u>

A + B → AB   **Redox (oxidation-reduction) <u>may</u> occur.**

Examples: $H_2 + N_2 →$ $2NH_3$

$H_2 + O_2 →$ $2H_2O$

$SO_3 + H_2O →$ H_2SO4

$P_4O_{10} + 6H_2O →$ $4H_3PO_4$

$2Fe + 3Br_2 →$ $2FeBr_3$

<u>**Decomposition**</u>

AB → A + B Redox <u>may</u> occur.

Example: Hypothetically, reverse the above reactions. Include: $2HgO →$ $2Hg + O_2$

Example: Metal carbonate decomposition to the corresponding metal oxide and CO_2: Let "Y" = a metal ion.

The reaction is: "Y"$(CO_3)_x →$ «Y»$O_x + CO_2$

$CaCO_3 → CaO + CO_2$

$Na_2CO_3 → Na_2O + CO_2$

$Al_2(CO_3)_3 → Al_2O_3 + 3CO_2$

Example: Metal hydroxide decomposition to the corresponding metal oxide and H_2O: Let "Y" = a metal ion.

The reaction is: "Y"$(OH)_x$ «Y»$O_x + H_2O$

$2KOH → K_2O + H_2O$

$Cu(OH)_2 → CuO + H_2O$

$Mg(OH)_2 → MgO + H_2O$

Example: Metal chlorate decomposition to the corresponding metal chloride and O_2: Let "Y" = a metal ion.

The reaction is: "Y"$(ClO_3)_x$ → «Y»$Cl_x + O_2$

$2KClO_3$ → $2KCl + 3 O_2$

$Mg(ClO_3)_2$ → $MgCl_2 + 3 O_2$

$2Ga(ClO_3)_3$ → $2GaCl_3 + 9 O_2$

Example: Metal nitrate decomposition to the corresponding metal nitrite and O_2: Let "Y" = a metal ion.

The reaction is: "Y"$(NO_3)_x$ → "Y"$(NO_2)_x + O_2$

$Ba(NO_3)_2$ $Ba(NO_2)_2 + O_2$

$2NaNO_3$ → $2NaNO_2 + O_2$

$2Al(NO_3)_3$ → $2Al(NO_2)_3 + 3 O_2$

Single displacement/replacement (see "activity series" in your textbook for metal displacements)

A + BC → AC + B A single displacement reaction is actually a redox reaction.

Examples:

$Zn + 2HCl$ → $ZnCl_2 + H_2$

$2Na + CuCl_2$ → $2NaCl + Cu$

$Cu + 2AgBr$ → $CuBr_2 + 2Ag$

$Cl_2 + NaI$ → $NaCl + I_2$ (*The displacement can occur with anions as well as cations)

Double displacement/replacement

AD + BC → AC + BD

Examples: $NaCl_{(aq)} + AgNO_{3(aq)}$ → $AgCl_{(s)} + NaNO_{3(aq)}$

$3CaCl_{2(aq)} + 2Na_3PO_{4(aq)}$ → $Ca_3(PO_4)_{2(s)} + 6NaCl_{(aq)}$

$Fe(NO_3)_{3(aq)} + 3NaOH_{(aq)}$ → $Fe(OH)_{3(s)} + 3NaNO_{3(aq)}$

A double displacement reaction is often driven by the ultimate formation of a less soluble phase, such as a precipitate (solid), coming out of solution. However, it may not always yield a precipitate, but a molecular substance, such as liquid water or gaseous carbon dioxide, especially in the case of **acid-base** reactions:

$HCl_{(aq)} + NaOH_{(aq)} \rightarrow HOH_{(l)}$ (water) $+ NaCl_{(aq)}$. There are obviously many variations of this theme.

$$K_2CO_{3(aq)} + 2H^+_{(aq)} + 2Cl^-_{(aq)} \rightarrow H_2CO_{3(aq)} + 2KCl_{(aq)}$$

About the Author

Eric G. Chesloff received his Bachelor's degree in chemistry from Lafayette College and his Master's degree in chemistry from Indiana State University. He spent several years in the pharmaceutical industry as a research chemist in liquid chromatographic analysis. For more than 35 years, he has taught and tutored university level general and biochemistry at several colleges and universities, among them Villanova University and St. Joseph's University. During this period, he returned to school and earned a Ph.D. in higher science education with a specialty in inquiry learning. Since then, his interest has been in novel undergraduate chemistry teaching methods.

Dr. Chesloff's teaching success in the local college classroom ignited a passion to reach out to chemistry students globally online.

When not in the classroom, Dr. Chesloff's interests include traveling, gardening, reading, and jazz guitar. He and his wife, Janis, live in the Philadelphia area.

Index

covalent bonding 73
covalent character 75
cross-multiplication 8, 23

D
Dalton dance 12, 15, 22, 73, 79
Dalton's law 47
Dalton's theory 12
DeBroglie 63
density 16, 45, 80, 82, 87
diamagnetic 67
dilute 29, 32, 33
dilution 30, 31, 32
dimensional analysis 8

E
effective nuclear charge 70
Einstein 62
electrical neutrality 21, 93
electromagnetic radiation 61
electromagnetic spectrum 61
electromagnetic waves 61
electron affinity 72
electron configuration 60, 61, 70, 74, 76, 77
electron delocalization 78, 88
electron density 80, 82
electronegativity 74, 75, 83
electrons 4, 6, 12, 14, 19, 20, 60, 62, 63, 64, 65, 66, 67, 68, 69, 70,
     71, 72, 73, 74, 75, 76, 77, 78, 79, 80, 81, 82, 84, 85, 87, 89
electron sharing 73, 74, 75
element 12, 15, 21, 67, 68, 69, 72, 76
endothermic 51
enthalpy function 54
Ernest Rutherford 14
exothermic 51, 60
experimentation 5, 11

F
factor label method 8, 9, 23, 31, 34
feta cheese 11
formula writing 93
fundamental stability principle 77

G
gamma rays 14
gases 29, 42, 47, 48, 49, 53
grams 13, 16, 23, 24, 25, 26, 28, 30, 31, 45, 46, 47, 56, 57, 59

M

Marie Curie 14
mass 12, 13, 15, 16, 19, 22, 23, 24, 25, 27, 30, 31, 35, 45, 56, 64, 70
matter 2, 3, 11, 13, 15, 16, 22, 40, 49, 54, 56, 59, 61, 62, 64, 82
Mendeleev 70
metallic behavior 72
mixtures 15, 29
modern chemistry 11
modern periodic law 70
molarity 30, 31, 32, 33, 34, 36
molar mass 23, 45
mole conversion 34
molecular compounds 21
molecular geometry 80
molecular motion 49, 50
molecular orbitals 88
molecular shapes 80, 81
molecules 4, 6, 23, 43, 44, 45, 48, 49, 50, 55, 63, 76, 78, 83, 84
moles 23, 24, 25, 29, 30, 31, 32, 36, 42, 43, 46, 47, 48, 59
monatomic ion 91
Moseley 70
MO theory 88

N

nearest noble gas neighbor 70, 74, 76
net nuclear charge 70
neutrons 12, 19
noble gas configuration 6, 69
noble gas rule 79
non-metals 21, 69, 74
nucleus 12, 19, 62, 65, 66, 70

O

orbitals 65, 66, 67, 69, 77, 84, 85, 86, 87, 88
organic chemistry 74, 78, 79, 88
outer electrons 6, 70
oxidation number 41
oxidation-reduction 39, 41, 95

P

paramagnetic 67
paramagnetism 88
particles 4, 14, 20, 21, 23, 48, 49, 60
Pauli exclusion principle 67
periodic table 12, 15, 21, 67, 68, 69, 70, 73
pi bonding 88
Planck's constant 62
polarity 74, 75, 83

stoichiometric 45
stoichiometry 23, 28, 30, 33, 34, 58
subatomic particles 4, 60
substances 12, 15, 22, 24, 39, 51, 74, 79
summary statement 5, 6, 15
sustained electron orbits 64

T

tetrahedron 82, 83, 84, 87
theoretical yield 26, 27, 29
theory 5, 6, 11, 12, 14, 48, 62, 80, 83, 85, 86, 88, 89
thermochemical 60
thermochemistry 49, 79
titration 33, 34

U

uncertainty principle 64
universal gas constant 44

V

valence bond theory 84
van der Waals equation 49
Voi-la! 87
volume 12, 16, 19, 30, 31, 32, 33, 34, 35, 43, 44,
        45, 46, 49, 54, 55, 56, 71, 81
VSEPR 80, 81, 82, 83, 84, 88
VSEPR theory 80

W

wavelengths 61
wave properties 64
Wilhelm Roentgen 14
work 29, 30, 41, 49, 50, 53, 54, 56
work function 53, 54
work (w) 53

9 798671 395297